I0470683

Sherwin-Williams Paint Warehouse Fire
Dayton, Ohio

With Supplement on
Sandoz Chemical Plant Fire
Basel, Switzerland

Investigated by: Tom D. Copeland
Philip Schaenman

This is Report 009 of the Major Fires Investigation Project conducted by TriData Corporation under contract EMW-86-C-2277 to the United States Fire Administration, Federal Emergency Management Agency.

 Homeland Security

Department of Homeland Security
United States Fire Administration
National Fire Data Center

U.S. Fire Administration Fire Investigations Program

The U.S. Fire Administration develops reports on selected major fires throughout the country. The fires usually involve multiple deaths or a large loss of property. But the primary criterion for deciding to do a report is whether it will result in significant "lessons learned." In some cases these lessons bring to light new knowledge about fire--the effect of building construction or contents, human behavior in fire, etc. In other cases, the lessons are not new but are serious enough to highlight once again, with yet another fire tragedy report. In some cases, special reports are developed to discuss events, drills, or new technologies which are of interest to the fire service.

The reports are sent to fire magazines and are distributed at National and Regional fire meetings. The International Association of Fire Chiefs assists the USFA in disseminating the findings throughout the fire service. On a continuing basis the reports are available on request from the USFA; announcements of their availability are published widely in fire journals and newsletters.

This body of work provides detailed information on the nature of the fire problem for policymakers who must decide on allocations of resources between fire and other pressing problems, and within the fire service to improve codes and code enforcement, training, public fire education, building technology, and other related areas.

The Fire Administration, which has no regulatory authority, sends an experienced fire investigator into a community after a major incident only after having conferred with the local fire authorities to insure that the assistance and presence of the USFA would be supportive and would in no way interfere with any review of the incident they are themselves conducting. The intent is not to arrive during the event or even immediately after, but rather after the dust settles, so that a complete and objective review of all the important aspects of the incident can be made. Local authorities review the USFA's report while it is in draft. The USFA investigator or team is available to local authorities should they wish to request technical assistance for their own investigation.

For additional copies of this report write to the U.S. Fire Administration, 16825 South Seton Avenue, Emmitsburg, Maryland 21727. The report is available on the Administration's Web site at http://www.usfa.dhs.gov/

U.S. Fire Administration

Mission Statement

As an entity of the Department of Homeland Security, the mission of the USFA is to reduce life and economic losses due to fire and related emergencies, through leadership, advocacy, coordination, and support. We serve the Nation independently, in coordination with other Federal agencies, and in partnership with fire protection and emergency service communities. With a commitment to excellence, we provide public education, training, technology, and data initiatives.

TABLE OF CONTENTS

SHERWIN-WILLIAMS PAINT WAREHOUSE FIRE
Dayton, Ohio
May 27, 1987

Local Contacts: Chief Glenn Alexander
District Chief Paul Hemmeter
District Chief Gary Douglas
District Chief Jim Dunham
Dayton Fire Department
300 N. Main Street
Dayton, Ohio 45402
(513) 443-4200
Mark Chubb, Plans Examiner
Douglas Hall, Water
Department

OVERVIEW

The Dayton, Ohio Fire Department avoided a double disaster by not attempting to extinguish a massive fire in a paint warehouse. The fire started on May 27, 1987, and completely destroyed the Sherwin-Williams Paint Warehouse. The dollar loss was $32 million, but only one employee was seriously injured and one firefighter sprained his leg. The noncombustible, sprinklered warehouse contained over 1.5 million gallons of paints and other products and was located over the aquifer from which wells provided the water supply for about one-third of the area's 400,000 people. Uncontained water and chemical run-off from firefighting could have contaminated this water supply and caused a greater loss than the fire itself, as occurred in Switzerland after the Sandoz Chemical Warehouse fire in 1986 contaminated the Rhine.

1

SUMMARY OF KEY ISSUES

Issues	Comments
Cause	Spark from lift truck ignited spilled contents of cans.
Industrial Firefighting	Ineffective; could not stop initial fire.
Sprinkler Systems	Overwhelmed
	Helped save office building adjoining warehouse.
	Question whether standard is adequate for this occupancy.
Firewall	Failed.
Incident Command	Highly effective.
	Quick, appropriate decision by first arriving district chief not to let crews in close.
	Courageous decision by Chief Glenn Alexander not to vigorously attack fire with water precluded environmental disaster to aquifer and city water supply.
Insurance	To be paid despite decision not to extinguish.
Municipal Liability	None apparent in this case but cannot generalize.
Prefire Plan	Did not address threat to aquifer.
Environmental Pollution	Apparently little to none.
	Tradeoff of air versus water pollution considered early, unlike Swiss Sandoz fire.
	Air and water experts on the scene early.
Firefighter Safety	A major factor in decision not to fight in close.
	Only one firefighter slightly injured in four-day fire.
Employee Safety	Difficulty in accounting for employees who escaped.
	Only one employee seriously injured.

THE FACILITY

The Sherwin-Williams Paint Warehouse was a large one-story building with an attached office building. An adjoining roofed-over area was used for drum storage. Trailers, some loaded, were located at the docks and within the fenced-in area of the facility. Direct exposure to properties outside the storage complex was minimal. (For a diagram of the site see Figure 1 in Appendix B.)

The main building's outside walls were of "tilt-up" concrete construction with a fire division wall of similar construction dividing the warehouse into two areas. The roof was supported by unprotected steel bar joists and columns. The facility was built in 1977 and contained about 190,000 square feet with walls about 33 feet high.

The building had a sprinkler system and diesel fire pump. The 2,500 gpm fire pump was located in a small detached building. Fire department connections to supplement the system were located on the warehouse side of the pumphouse.

The pump was supplied by a 12-inch line connected to a 16-inch public water main. The pump supplied a 10-inch loop around the site with connections to various sprinkler risers, external exposure sprinklers along the outside wall at the drum storage, fixed monitors for the drum storage area, and private hydrants. Water flow alarms were monitored by a central station. The ceiling sprinklers had a reported design density of 0.32 gpm per square foot over 4,000 square feet.

The warehouse contained over 1.5 million gallons of paints and related flammable liquids mostly in small containers up to 5-gallon sizes. There was also considerable storage of aerosol cans. Drums were stored outside under the canopy area. Trailers contained additional products. The warehouse had extensive rack type storage, up to four tiers of pallets high. Approximately 30 employees were working at the time of the fire.

THE WATER SYSTEM

The public water supply for Dayton, Ohio and surrounding areas is drawn partly from an aquifer over which the Sherwin-Williams Paint Warehouse was built. The warehouse was within a major well field with wells on and adjacent to the site. (See Appendix D for map of wells.)

Chemicals can migrate down from the surface and contaminate the water. This was a concern when the industrial park where the warehouse was located was first developed. No detailed plans had been made by the fire department to deal with a threat to the water supply, although the threat was anticipated and discussed several years before the fire.

THE FIRE

During the evening of May 27, 1987, a workman using a motorized lift truck accidentally knocked over and spilled several cans of flammable liquid in the east portion of the warehouse. The liquid probably was ignited by a spark from the electric motor of the truck. The lift truck operator received serious burns and was helped by other employees to put out his flames and escape. The employees quickly decided that the fire was beyond their ability to extinguish, and all evacuated without further injuries. They credited company training for knowing what to do.

The Dayton Fire Department received the alarm automatically from the fire alarm central station when the plant's sprinklers activated, along with many calls from the public and employees. Three engines, one truck, and a district chief were dispatched at 2107. While responding, the district chief quickly decided to request a second alarm at 2108. This resulted in four additional engines, two additional trucks, and another district chief. The first engine on the scene reported complete building involvement at 2113. The first-in district chief requested a third alarm as he arrived on the scene at 2114.

A total of 84 Fire Department personnel responded with ten engines, five trucks, and twelve other vehicles. Most of the equipment was staged and not actually used.

The fire was through the roof, and the east half of the warehouse was totally involved when the first units arrived. Aerosol cans were raining on the crews and hurtling for distances. The initial concerns were for protecting the drum storage, office, and other exposures, and for fire fighter safety. The district chief quickly decided not to let his fire fighters hook up to the sprinkler supply of the pumphouse, which was close to the wall of the warehouse and exposed to intense heat, danger from exploding and the hurtling cans, and the threat of a wall collapse. Also, it was thought that the sprinkler system was probably compromised already. The initial attack was directed at the office, drum storage, and loaded trailers in the docking and parking areas.

The fire spread past the fire wall and was involving the west half of the warehouse before plans could be implemented to cut the fire off at the wall. At 2132, the dispatcher was advised that the building would be a total loss. (The fire ultimately did destroy the main warehouse, and 17 trailers also were heavily or totally damaged. The office, its records, and the outdoor drum storage were saved.)

The warehouse sprinkler system was overwhelmed almost immediately and had little impact on the main fire. Water was observed discharging from broken sprinkler piping early in the fire. Sprinklers did apparently contribute to saving the office building. The fire pump supplying the sprinklers was manually shut down at about 2235 or 2245.

Automatically opening plastic vents almost directly over the incipient fire operated early and may have released heat that otherwise would have built up quickly and caused more sprinkler heads to go off, suggested Dayton Chief Glenn Alexander. In any event, the sprinkler system design was not adequate for controlling this fire.

Chief Alexander assumed incident command early on and ordered that no water be applied to the fire in the warehouse. Water was to be used only to protect exposures and to extinguish fires where the run off could be monitored on paved areas. Because of his concern about the water supply well field, Chief Alexander called the director of the water department prior to responding and requested the director to meet him at the scene.

The contents of the warehouse, the problem of air pollution versus water pollution, and other conditions were considered. In addition to local water authorities, state air and water pollution experts were on the scene the first day. It was agreed that it would be best not to increase the threat to the water system by applying water to the warehouse fire. The smoke was not causing a direct life safety exposure and was described as being similar in hazard to diesel exhaust.

Douglas Hall, Environmental Protection Manager for the Dayton Water Department, said the early decision not to put water on this fire was clear cut. The decision was made easier due to the knowledge that there were no highly toxic materials present and no residential areas close-by. If it had not been for the threat to the underground water supply, they probably would have recommended applying water to the fire and controlling run-off. Although the fire department avoided applying much water to the warehouse itself, the sprinkler system's 2,500-gpm fire pump did operate for a while and there also was runoff from the initial attack. Thus some contaminated water reached the nearby Miami River. On day one of the fire, paint solvents were skimmed and pumped from the river. A water containment dike was started on the fourth day of the fire in preparation for the final extinguishment operation. Since the fire, numerous test and recovery wells have been dug. There is some ground contamination, but the contamination apparently is being managed and has not yet affected the water supply.

The fire was reported contained at 0012 hours on May 28, 1987, but not under control until 1004 hours on June 2, 1987, almost six days after it started.

CODE COMPLIANCE

The warehouse was in compliance with the Ohio Building Code, as best could be determined. That code allowed unlimited space in the warehouse so long as it was fully sprinklered. No performance standards or guidelines are provided in the code as to what constitutes adequate sprinklering for such a facility.

There is some question as to what would be the most cost-effective fire protection design that would comply with the code for a warehouse such as this, full of highly flammable liquids.

One alternative is to subdivide the building into many compartments, each with a high flow sprinkler system--practically like preparing to flood a magazine compartment on a ship. Another is to build large open spaces with few firewalls, lower flow sprinkler system that may handle some fires (such as cardboard cartons igniting or a small spill) but not one such as occurred. A third approach-- not in compliance with most codes--is to build a lightweight, "disposable" building or shed (such as was involved in the Swiss Sandoz Chemical Warehouse fire discussed in the supplement) but built in a safe, remote environment (unlike the Sandoz fire) where it could be allowed to burn. Containment ponds and dams can be built-in to catch water runoff if water pollution could be a problem. Further fire protection studies are needed to examine the various other options for such warehouses.

LIABILITY

Was the fire department exposed to liability suits by deciding not to extinguish the fire? Do insurance companies have to pay insurance in these cases? In this case, the answers were largely moot because the building was judged to be on its way to becoming a total loss when the decision was made, and the insurance company tentatively agreed to pay the loss without protest. Here is a synopsis of the liability situation, but this is no substitute for seeking legal advice for your own area.

1. The fire department has very broad rights to take action in an emergency. It even can destroy property to save other property; for example, during the San Francisco earthquake of 1906, rows of buildings were dynamited to provide a fire break to stop the fire. The owners, however, can seek indemnification from the city for destroying their property. The insurance company also can sue.

2. The fire department and the city can be liable for negligence (unreasonable actions) unless state statutes limit that liability or unless the cities have "sovereign immunity." Most states no longer recognize the latter. States vary in their liability laws. Liability questions need to be answered state by state.

3. Most states have laws that limit the liability of a fire department for negligence or poor judgment. The limit may be zero as in California. In Minnesota, it is $300,000 or the limit of negligence insurance it carries, whichever is greater.

4. If a fire department decides to let a building burn that could have been saved in order to protect a greater loss to the community, they could be sued by any party--the property owner, insurance company, public, etc., if found negligent, then 2) applies. If not, they are home free--except for legal expenses.

5. The insurance company usually has to pay for the loss regardless of the fire department actions. Although the insurance company reasonably expects the fire department to fight a fire if their insured property burned, it is their hard luck if the fire department does not fight the fire. However, knowing this possibility exists may cause premiums to go up for properties that could cause pollution if they burn. Also, the insurance company might claim that the building owner did not reveal all hazards, and try to hold back a part or much of the insurance, or delay payment.

6. The insurance company might have to pay for environmental damage up to the limit of the policy as part of fire losses. It depends on the details of the policy and/or the details of the situation.

7. This is all on the edge of a new area legally, environmentally, ethically, and from fire fighting points of view. It needs further exploration. Fire departments should discuss the issue with local city attorneys. The fire department, while it generally appears to be safe, may fall into some loophole or have an adverse interpretation of the law in light of the new circumstances.

LESSONS LEARNED

1. **Risk Management**--The most important lesson learned because of this fire is not simply that some fires should be allowed to burn but that the consequences of all actions and "inactions" must be knowledgeably considered.

    Chief Alexander describes today's fire chiefs as "risk managers." This is a good application of the term and broadens a chief's role and responsibility. Risk management of such fires as this involves the consideration of:

    > Characteristics of materials and chemicals involved
    > Air versus water pollution
    > Wind and weather conditions
    > Capability to extinguish or control the fire;
    > Ability to contain run-off
    > Short-term versus delayed hazards
    > Life safety and property exposure
    > Evacuation problems.

2. **Water Pollution**--In this fire, the decision not to apply water to the warehouse fire resulted in far less contamination to the ground water and little if any difference in property loss. State and local air and water pollution experts were brought to the scene early to consult. The Swiss Sandoz Chemical Plant fire, which polluted the Rhine, demonstrated what can happen when water run off is not considered. (A summary of that fire is presented below.) However, it may not always be possible to allow such fires to burn when there is a high exposure hazard or an air pollution problem. Applying water to avoid a fire or air pollution catastrophe may be the lesser evil at times; it depends on the situation. Containment of water runoff should be a consideration both in prefire planning and in planning fire protection systems for a structure or complex which has significant amounts of hazardous materials.

3. **Insurance and Law Suits**--According to Chief Alexander, the insurance company for the warehouse said it will not sue the Fire Department for not applying water to extinguish the warehouse fire. A key factor here was that the warehouse building was essentially a total loss at the time the decision to stop applying water was made by the Chief. If the building could have been saved, the same finding might not have been made.

    The liability of the city might have been much greater if the water supply had been damaged, let alone the adverse local and national publicity that would surely have ensued. Nevertheless, fire departments that plan ahead of time to let a fire burn because of environmental considerations should one occur on a particular property need to discuss that possibility beforehand with the city attorney and the property owner as part of pre-fire planning.

4. **Sprinkler Systems**--Two important fire safety features provided in this warehouse failed: the sprinkler system and the firewall. Apparently, the sprinkler system was quickly overwhelmed and could not provide the necessary water density. This may have been affected by the venting system releasing heat directly above the incipient fire and stopping heat build-up that might have triggered other heads quickly. The system was considered in compliance with current standards and the Ohio codes; Chief Alexander has called for a reexamination of the standards for such high-risk occupancies.

   Turning off the fire pump and then the water supply to the warehouse was a calculated risk primarily in regard to the office area. The warehouse was not being affected by the sprinkler system, but the office area had been saved to that point by several heads that operated. It was felt that the threat to the aquifer outweighed the potential loss of the office building. However, after the fire, the office building was found to have received very little damage.

5. **Firewall**--The fire wall did not withstand the rapid fire build-up and intense exposure. A hole developed in it. Chief Alexander stated that the fire doors in the firewall did close except in one case where only a door on one side of the wall closed. The opposite door was jammed by debris. The opening was protected by the door from one side. Again, standards for firewalls in such high-risk facilities need to be reexamined.

6. **Firefighter Safety**--It was remarkable and a tribute to Dayton's Incident Command and fire fighters that they sustained only one minor injury in the course of this fire. (It was a strained leg from lifting hose.) As it should be, fire fighter safety was considered right from the early decisions not to supply the sprinkler system and to pull units well away from the walls.

7. **Land Development Decision**--The threat to the aquifers would not have existed if the paint storage facility had not been allowed to be built amidst the water well field. The city had had second thoughts about allowing development on this land and had stopped the full development originally planned. Environmental impacts need to be and often are a major factor in land development decisions. Potential impacts from fires are not always considered in these studies, and fire departments should try to make sure that they are where appropriate.

8. **Employee Training**--A Sherwin-Williams employee accidentally started the fire. Whether such fires can be totally prevented is debatable, and may not be economically feasible (e.g.; not allowing equipment that can produce sparks anywhere near flammable liquids should there be breakage and a spill.)

   The employees were well trained to evacuate quickly and rendezvous, and to extinguish the flames on the clothes of the lift truck driver. However, the place they were to rendezvous at was being barraged by exploding canisters, so the employees did not stay together, and a head count could not be taken to ensure all had escaped. Unnecessary time was spent tracking them down and ensuring that no one was missing. Employees should be instructed to go to a meeting place that, as far as can be determined in advance, will be safe and/or to check in after a disaster occurs.

9. **Incident Reporting**--Training is needed to ensure consistent and accurate reporting of fire incidents. In this case, for example, the sprinkler systems operated, but were not so reported. With understandable intent, the fire officer filling out the report noted that the sprinkler performance was "other--not described above" because the sprinklers were overwhelmed. In analyzing this data across many fires, this clear case of sprinkler system failure would not have been counted.

10. **Incident Command**--The fire was a near textbook example of the use of a good incident command system. Higher level chiefs smoothly took over as incident commander as the alarms built up. There were no major communications problems. The dispatchers did a good job of coping with a large number of callers who reported the fire or asked about its risk. The incident command helped hold casualties and losses down.

*****

Because historically the objective of firefighters has been to extinguish hostile fires, it is hoped that a review of this fire will make it apparent that risk management is a higher objective and that it is necessary to make decisions that result in the lowest possible immediate and long-term loss even if that means letting the fire burn.

It is instructive to compare the results of the Sherwin-Williams fire with the 1986 fire in the Sandoz Chemical Plant in Basel, Switzerland, which is described in the following supplement.

SUPPLEMENT

Pollution of Rhine River Due to Runoff from Sandoz Chemical Plant Fire in Basel, Switzerland[1]

Thirty tons of toxic material washed into the Rhine River with water firefighters used to fight a warehouse blaze at a riverside Sandoz Chemical Plant and Storage Facility near Basel, Switzerland in the early morning hours of November 1, 1986.

By the time the chemicals, mostly pesticides, had traveled 500 miles down the winding scenic river, half a million fish were dead, several municipal water supplies were contaminated, and the Rhine's ecosystem was badly damaged with virtually all marine life and a large proportion of microorganisms wiped out.

The approximately 25-mile-long chemical slick drifted slowly downstream from the

Swiss border to the North Sea. It contained about 30 tons of insecticides, herbicides, and mercury-containing pesticides, and threatened the North Sea's winter cod harvest. Environmental groups called for a boycott of Sandoz products.

In the weeks following the fire, citizen protest rallies occurred, the Swiss government as well as Sandoz Corporation received damage claims from other countries, and Switzerland had to respond to strong criticism for its handling of the emergency from France, West Germany, the Netherlands, Luxemburg, and the Common Market Commission.

THE FACILITY

The warehouse where the fire started was built in 1967. It was part of a large Sandoz chemical complex in Schweizerhalle, a small community six miles east of Basel on the Rhine's left bank. The warehouse was about 295 feet long by 82 feet wide, with an adjoining second half another 82 feet wide separated from the first by a wall down the length of the building. It had no sprinklers because the risk of a fire was considered low. The building in effect was a light shed intended to provide shelter from rain and extremes of temperature, rather than being a solid warehouse. Its height ranged from 26 feet to a peak of 39 feet.

The half of the building where the fire started was stacked with about 1,250 tons of chemicals in barrels four pallets high, somewhat like the Sherwin-Williams storage. The chemicals stored were mainly flammable liquids, including pesticides, fungicides, and herbicides, some with 30 degrees C flashpoint. Among these were phosphoric acid and organic mercury compounds. Among additional raw materials present were ferric ferrocyanide, which may have been a key factor in the ignition sequence. The other half (82 foot width) of the building had mostly harmless chemicals.

---

[1]The following sources were used in this supplement: Associated Press stories following the fire; a presentation by Hans Wackerlig, Fire Prevention Service, Zurich Switzerland at the NFPA Fall Meeting, Portland, Oregon, November, 1987; personal discussion between Wackerlig and Philip Schaenman, November, 1987, "The Lessons Learned From the Sandoz Fire," Hans Wackerlig, 1987.

THE INCIDENT

In response to simultaneous reports by a police highway patrol alarm and the plant night watchman at 0019 on November 1, 1986, three Sandoz plant brigade fire fighters and the chief responded to the warehouse. Flames were shooting from the roof when the fire was first noticed. Upon arrival, the chief immediately realized that he could not cope with the situation alone and called for an all-out alarm. By 0045, 200 fire fighters were in action at the scene.

The cause of the fire has not been positively determined. It might have been started by the ignition of the ferric (ferrocyanide in the warehouse) by a butane-powered machine used to shrink-package chemicals in plastic films. The ferrocyanide was being packaged earlier in the day. This chemical has the insidious property--discovered only after the fire--of smoldering without releasing any smoke or odor, and then suddenly breaking into almost explosive burning. Ironically, the packaging of the chemicals was started by a zealous employee who wanted to tidy up the storage. While this seems the likely cause, arson has not been ruled out.

Because the fire was not discovered until it was already large and being fed by a warehouse full of highly flammable chemicals, it was accepted from the start that the warehouse would be a total loss. Attention was focused on stopping exposure fires, no mean task since barrels of flammable chemicals were hurtling through the air. At first the fire fighting was defensive, but then the chief decided to try to extinguish the fire with massive amounts of water to stop the fire spread and avoid a catastrophe to the nearby city and three major chemical complexes nearby. There also was a great deal of attention given to the risk from the possibility of toxic clouds of gases being generated and whether the nearby populations in Switzerland, France, and Germany would have to be evacuated.

More than 3,000 gallons of water a minute was being pumped from the Rhine to fight the fire and keep it away from neighboring warehouses and outdoor storage. The peak pumping rate reached 8,000 gpm.

A 12,000-gallon catch basin into which both water and chemicals collected began overflowing into the river. Flames rose to 200 feet above the warehouse. Steel drums of chemicals exploded like bombs in the intense heat; gas and smoke spread towards the outskirts of Basel. At 3:30 a.m., a hastily convened regional crisis staff declared an emergency. No evacuation was needed. The fumes were not thought to be toxic but included mercaptans, one of the most malodorous chemicals known to man, one which causes people to feel sick and fearful that they are being poisoned. Area sirens were sounded and radio announcements urged the population to close windows and stay indoors. (Many sirens were down for routine maintenance and could not be used.)

Public transport into the area was halted, and gas masks were prepared at the civil defense arsenal. Officials ended the emergency 90 minutes later when readings showed no dangerous concentrations of toxicity in the air. No one was hurt and calm returned to Basel.

But slowly the massive run-off began moving down the Rhine. The management of the Sandoz Chemical Plant sent telexes to all municipal water systems along the 520 miles of river between Basel and the Dutch North Sea port of Rotterdam, urging tests on pollution levels. It listed eight toxic chemicals, most of them used in pesticides, that may have washed into the river. A number of West German water systems were shut down and populations supplied with drinking water by tank trucks. Dutch authorities ordered services closed to keep contaminated water out of Rhine estuaries. Most of the water applied to the fire flowed off through storm drains to the Rhine. By 4:30 a.m. the fire was under control. Incredibly, however, no serious thought seems to have been given to the potential water pollution. It was not the immediate and present danger.

AFTERMATH

The full extent of the ecological damage was not evident for a few days. West Germany's Parliament was told that half a million fish were killed and aquatic life had ended in large stretches of the river. Fisheries officials said new fish for breeding probably would not be introduced for several years and it might be ten years before the river recovers. Heavy metal pollutants which sank to the bottom continued to be stirred up, sending out additional waves of pollution. It was especially tragic because the fishlife had only recently returned to the Rhine after massive clean up operations in the previous years.

But the predictions were overly pessimistic: life appears to be returning to the river today, one year after the disaster caused by man.

The warehouse where the fire originated and its adjoining twin (the double width) were destroyed, but none of the others nearby nor the open air storage were destroyed. Though the Rhine was seriously polluted, it could have been a much worse, more toxic fire had it spread further to other nearby warehouses. There were no injuries, though about 150 civilians and fire fighters were given blood tests to see if they had elevated mercury levels or other problems of the blood. They are still being monitored to see if any long-term effects appear, but none have so far.

LESSONS LEARNED

The Swiss View

There were many lessons learned from this fire. The largest and most important lesson was that water pollution has to be considered in fire fighting, along with other environmental factors. Fires may need to be allowed to burn. "Whereas fire prevention and environmental protection were previously regarded as two completely independent fields with some slight overlap of common interest... the interface is now realized to be much more important than had been assumed." However, it is still thought that it was necessary to extinguish the fire to keep it from spreading.

How to contain the water runoff needs to be considered in planning fire protection, especially where toxic chemicals or things that produce toxic chemicals when burned are present.

A third major lesson was the need for greater security around warehouses. Fifty percent of warehouse fires in Switzerland are from arson, they report. Even if this fire had not been started intentionally, the potential was there. Plant security was deemed far too lax. This was especially so for a chemical industry under political attack, as was Sandoz at the time of the fire.

A fourth lesson was the need to better label toxic substances as to fire hazard, personal hazard, and environmental hazard. A set of new symbols has been proposed for Common Market use by a working group organized to study the fire's implications.

A fifth lesson was that fire prevention and built-in fire protection for chemical warehouses need to be rethought. Present planning has proven inadequate.

APPENDICES

A. Dayton Fire Department Photographic Slides (with master file copy at U.S. Fire Administration).

B. Photographic Slides from Investigator (with master file copy).

C. Fire Incident Report

D. Map of Wells near Sherwin-Williams Plant (with master file copy).

E. Transcripts from Fire Department Telephone, Radio, and PA.

F. Newspaper Articles with Map of Site Reprinted with permission of the *Dayton Daily News* and *Journal Herald*. (Additional articles are with the master file copy at USFA.)

APPENDIX A

SHERWIN-WILLIAMS WAREHOUSE FIRE

Description of Slides

DAYTON FIRE DEPARTMENT PHOTOGRAPHIC SLIDES

1D. View of large flame mass from west (?) side of warehouse.

2D. Aerial view of fire scene on Friday after Wednesday night fire. View toward east with south side of building to right and west side to near left.

3D. Aerial view of fire scene on Friday after Wednesday night fire. View to west side of warehouse.

4D. Aerial view of fire scene on Friday after Wednesday night fire. Southwest corner of building in foreground. Note concrete tilt-up outside walls still standing at east end of south wall.

5D. View of remains after fire. Note paint cans and twisted steel columns.

6D. View of sprinkler head and pipe fallen amid debris of aerosol cans in area of fire origin.

7D. View of remains of fallen, reinforced concrete "firewall" after fire.

8D. View of north end of north/south firewall. Note large spalled hole through wall.

9D. View of water discharging from broken sprinkler risers on west side of firewall. (Note hole in firewall for reference.) Fire pump believed to have been shut off prior to this photograph and water flow is from bypass around pump.

10D. View of 55-gallon drum storage under canopy at east end of warehouse.

Appendix A (continued)

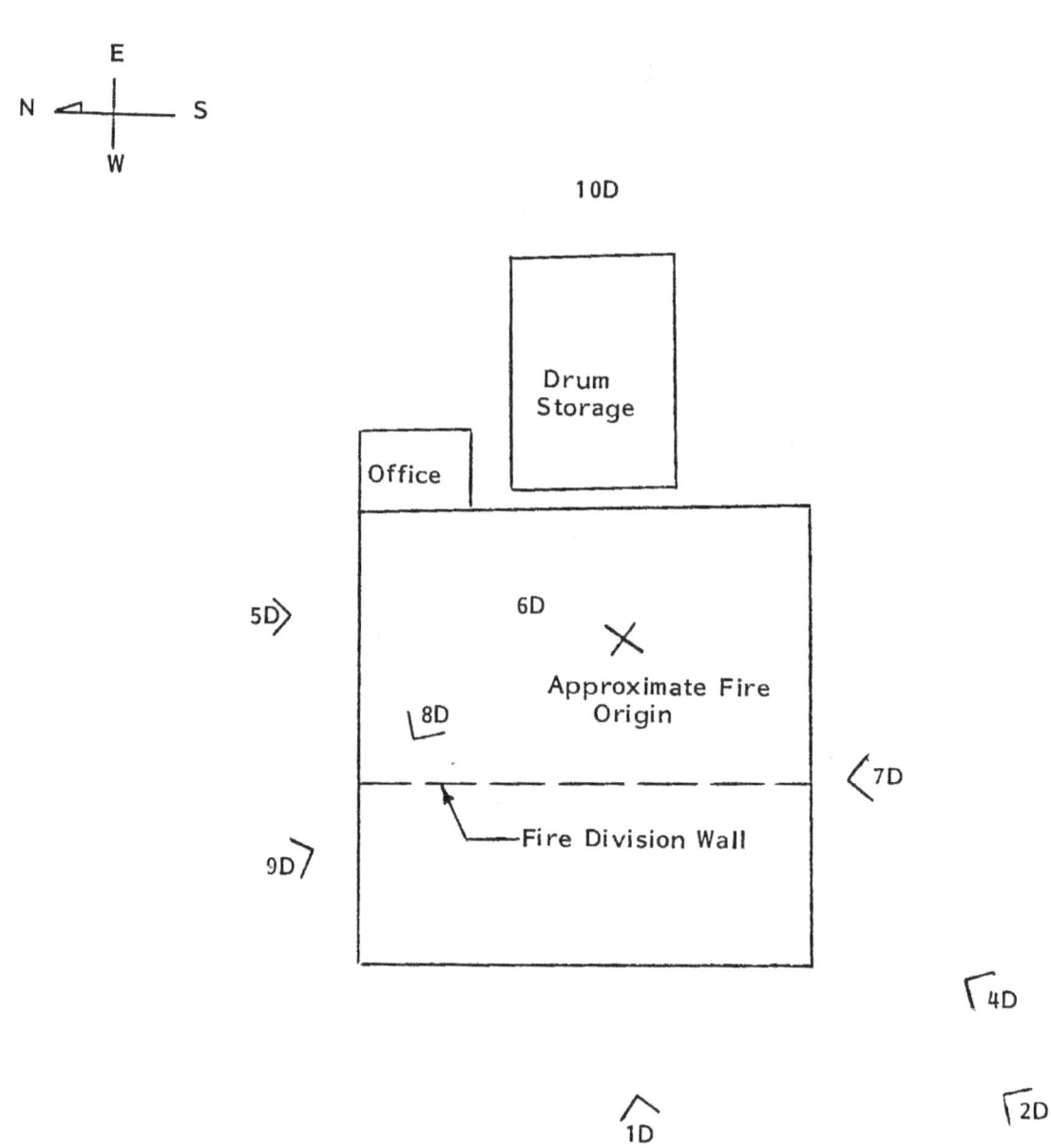

DAYTON FIRE DEPARTMENT PHOTOGRAPHIC SLIDES

Area or Direction of View of Slides
(Description Attached)

267T 1043	10-10-87
Sherwin-Williams Warehouse	
NTS	

APPENDIX B

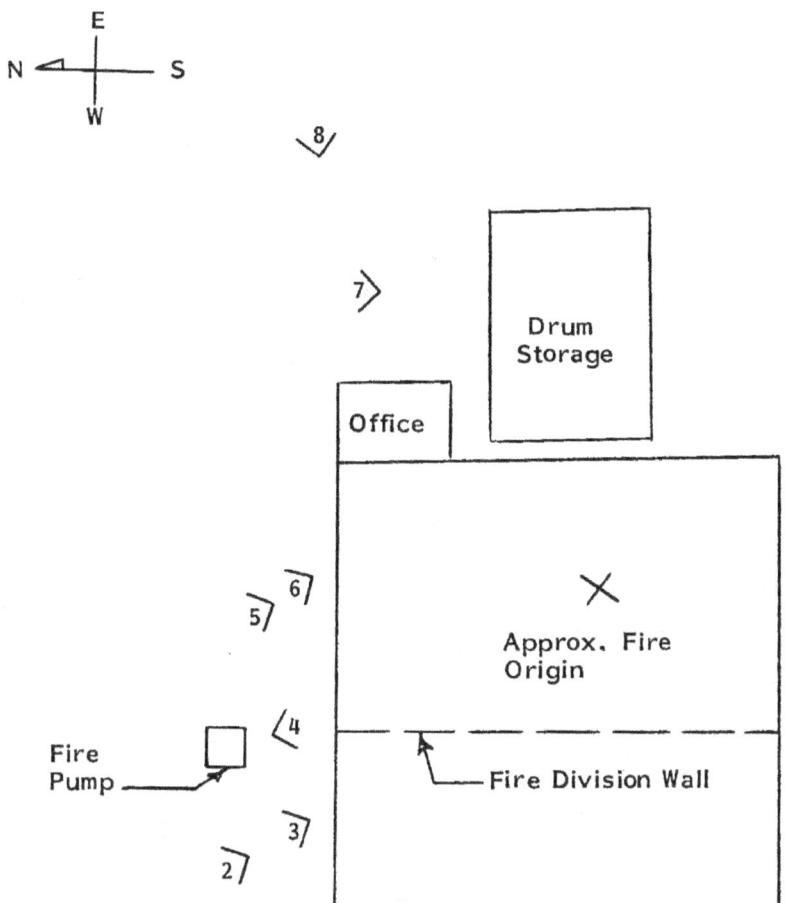

SLIDES:

#1 Diagram of Site with Information (Provided by Dayton Fire Dept.).

#2 View toward SE showing remains of "Fire Wall", small attached bldg.,
and east wall with office beyond in left background.

#3 Closeup of fire wall remains showing hole in wall. Only one panel remaining.

#4 View of south side of fire pump house. Pump and connections have been removed.

#5 View toward SW showing remains of dock and east wall.

#6 View toward SW showing closeup of east wall and openings into office area.

#7 View toward south showing remains of canopy where drums were stored.

#8 View toward west showing office area and remains of east wall of warehouse.

267T1043	8-2-87
Sherwin-Williams Wharehouse	
NTS	*DC*

17

APPENDIX C

OHIO FIRE INCIDENT REPORTING SYSTEM

INCIDENT REPORT

1 ☐ DELETE
2 ☐ CHANGE

Fire Department _____

A | FDID: 5 7 0 1 1 | INCIDENT NO: 2 5 4 4 | EXP: 6 1 5 | MO: 2 | DAY: 7 | YEAR: 8 7 | DAY OF WEEK: 4 | ALARM TIME: 2 1 0 7 | ARRIVAL TIME: 2 1 1 6 | TIME IN SERVICE: 1 9 0 4

DAY OF WEEK: 1-Sunday, 2-Monday, 3-Tuesday, 4-Wednesday, 5-Thursday, 6-Friday, 7-Sat

6/2/87

B — SITUATION FOUND (GRASS, TRASH FIRES, SHORT FORM): 11-Structure Fire, 12-Outside Dollar Loss, 13-Vehicle Fire, 14-Brush, grass, leaves, 15-Trash, Rubbish, 16-Explosion, No other fire, 17-Outside spill with fire, 22-Air Gas Rupture, 32-Emergency Medical call, 33-Locked-in trapped, 34-Search, 35-Extrication, 41-Spill, leak-No fire, 42-Explosive, Bomb removal, 43-Excessive Heat, 44-Power line down, 45-Arcing electric equipment, 46-Aircraft standby, 47-Chemical spill, 51-Lock-out, 52-Water removal, 53-Smoke odor removal, 54-Animal Rescue, 55-Assist Police, 56-Unauthorized burning, 57-Move-up, 59-Other service calls, 61-Smoke scare, 63-Controlled burn, 65-Steam, gas mistaken for smoke, 71-Malicious false, 72-Bomb Scare, 73-Alarm Malfunction, 74-Unintentional false, 99-Unclassified | Other ____ See page 17-19 | **1 1**

B ACTION TAKEN: 1-Extinguishment, 2-Rescue or Assistance, 3-Investigation only, 4-Remove Hazard, 5-Stand by, 6-Salvage, 7-Ambulance, 8-Fill in, Move up, 9-Not classified, 0-Undetermined | PROTECTED EXPOSURES | **9** | MUTUAL AID: 1-Rec'd, 2-Given, N/A | **N/A**

C FIXED PROPERTY USE (Occupancy) Pg 23-43: PAINT STORAGE **8 6 5** | IGNITION FACTOR Pg 44-45: FLAMMABLE/LIQUID SPILL **4 1**

D CORRECT ADDRESS (Up to maximum of 21 characters): 3671 DAYTON PARK DRIVE | ZIP CODE: 4 5 4 1 4 | CENSUS TRACT: 1 8 0 7

E OCCUPANT NAME (LAST, FIRST, MI): SHERWIN WILLIAMS | TELEPHONE: 236-5282 | ROOM or APT.

F OWNER NAME (LAST, FIRST, MI): SHERWIN WILLIAMS | ADDRESS: 3671 DAYTON PARK DR. | TELEPHONE: 236-5282

G METHOD OF ALARM: 1-Telephone direct, 2-Municipal alarm system, 3-Private alarm system, 4-Radio, 5-Verbal, 6-No alarm recd, 7-Tie-line (911), 8-Voice signal municipal alarm signal, 9-Not classified above, 0-Undetermined or not reported | **3** | CO. INSPECTION DISTRICT: 1 2 | SHIFT: 2 | NO. ALARMS: 3

H NO. FIRE SERVICE PERSONNEL RESPONDED: 8 4 | NO. ENGINES RESPONDED: 1 0 | NO. AERIAL APPARATUS RESPONDED: 5 | NO. OTHER VEHICLES RESPONDED: 1 2

I (COMPLETE IF CASUALTY) NUMBER OF INJURIES — FIRE SERVICE: 1 | OTHER: 1 | NUMBER OF FATALITIES — FIRE SERVICE: 0 | OTHER: 0

J COMPLEX Pg 61-62: 9 8 | MOBILE PROPERTY TYPE Pg. 63-65 (Complete Line S) NA = 08: 9 8

K AREA OF FIRE ORIGIN Pg 67-68: 4 1 | EQUIPMENT INVOLVED IN IGNITION Pg 71-72 (Complete Line T) 98: 9 9

L FORM OF HEAT IGNITION Pg 74-76: 2 6 | TYPE OF MATERIAL IGNITED Pg 78-79: 2 1 | FORM OF MATERIAL IGNITED Pg 80-81: 5 7

M METHOD OF EXTINGUISHMENT: 1-Self extinguished, 2-Make shift aids, 3-Portable extinguisher, 4-Automatic ext system, 5-Pre-connect hose/tank only, 6-Pre-connect hose/hydrant draft standpipe, 7-Hand-laid hose/hydrant draft standpipe, 8-Master stream device, 9-Not classified above, 0-Undetermined or not reported | **1** | LEVEL OF FIRE ORIGIN: 1-Grade level to 9 ft, 2-10 to 19 feet, 3-20 to 29 feet, 4-30 to 49 feet, 5-50 to 70 feet, 6-Over 70 feet, 7-Objects in flight, 8-Below ground level, 9-Not classified above, 0-Undetermined | **1** | ESTIMATED TOTAL DOLLAR LOSS: 32.4 million | 3 2 4 0 0 0 0 0

N Number of Stories: 1-1 story, 2-2 story, 3-3 to 4 stories, 4-5 to 6 stories, 5-7 to 12 stories, 6-13 to 24 stories, 7-25 to 49 stories, 8-50 stories or more, 0-Number of stories undetermined or not reported | **1** | CONSTRUCTION TYPE: 1-Fire resistive, 2-Heavy timber, 3-Protected noncombustible, 4-Unprotected noncombustible, 5-Protected ordinary, 6-Unprotected ordinary, 7-Protected wood frame, 8-Unprotected wood frame, 9-Not classified above, 0-Undetermined or not reported | **4**

O EXTENT OF DAMAGE: Flame-Smoke — Confined to the object of origin (1,1), Confined to part of room or area of origin (2,2), Confined to room of origin (3,3), Confined to the fire-rated comp of origin (4,4), Confined to floor of origin (5,5), Confined to structure of origin (6,6), Extended beyond structure of origin (7,7), No damage of this type (N/A), Undetermined or not reported (0,0) | FLAME: 7 | SMOKE: 7 | **P** DETECTOR PERFORMANCE: 1-Det in room or space of fire origin - oper, 2-Det not in rm or space of fire origin - oper, 3-Det in rm or space of fire origin - no oper, 4-Det not in rm or space of fire origin - not oper, 5-Det in rm or space of fire origin but fire too small to oper, 9-Not classified above, 0-Undetermined or not reported, 8-No detectors present (N/A) | **1** | SPRINKLER PERFORMANCE: 1-Equipment operated, 2-Equipment should have operated - did not, 3-Equipment oper but fire too small to oper, 9-Not classified above, 0-Undetermined or not reported, 8-No equipment present (N/A) | OVERPOWERED SYSTEM | **9**

Q TYPE OF MATERIAL GENERATING MOST SMOKE Pg 103-104 (IF SMOKE SPREAD BEYOND ROOM OF ORIGIN): 2 1 | AVENUE OF SMOKE TRAVEL: 1-Air handling duct, 2-Corridor, 3-Elevator shaft, 4-Stairwell, 5-Opening on construction, 6-Utility opening in wall, 7-Utility opening in floor, 8-No avenue of smoke travel (N/A), 0-Undetermined or not reported, 9-Not classified above | ROOF | **9**

R FORM OF MATERIAL GENERATING MOST SMOKE Pg 108-109: 5 7

S IF MOBILE PROPERTY | YEAR | MAKE | MODEL | SERIAL NO. | LICENSE NO.

T IF EQUIPMENT INVOLVED IN IGNITION: FORK LIFT | YEAR: UNK | MAKE: UNK | MODEL: UNK | SERIAL NO.: UNK

U MEMBER MAKING REPORT: F.F. Michael Kenny S/ | DATE | OFFICER IN CHARGE (if different): D.C. Gary L. Douglas | DATE: 6-3-87

Remarks ___ See attached narrative sheet ___

☐ check if remarks continued on back

COM 8013

18

Appendix C (continued)

OHIO FIRE INCIDENT REPORTING SYSTEM

INCIDENT REPORT

NFIRS-1

1 ☐ DELETE
2 ☐ CHANGE

Fire Department _____

FDID	INCIDENT NO	EXP.	MO.	DAY	YEAR	DAY OF WEEK		ALARM TIME	ARRIVAL TIME	TIME IN SERVICE
5 7 0 1 1	25 24 1		1	0 5	8 7	1-Sunday 4-Wednesday 7-Sat / 2-Monday 5-Thursday / 3-Tuesday 6-Friday	4	21 07	21 16	07 00

SITUATION FOUND
11-Structure Fire 22-Air Gas Rupture 44-Power line down 55-Assist Police 72-Bomb Scare
12-Outside Dollar Loss 32-Emergency Medical call 45-Arcing electric equipment 56-Unauthorized burning 73-Alarm Malfunction
13-Vehicle Fire 33-Locked-in trapped 46-Aircraft standby 57-Move-up 74-Unintentional false
14-Brush, grass, leaves 34-Search 47-Chemical spill 59-Other service calls 99-Unclassified
15-Trash, Rubbish 35-Extrication 51-Lock-out 61-Smoke scare Other _____
16-Explosion, No after fire 41-Spill, leak-No fire 52-Water removal 63-Controlled burn See page 17-19
17-Outside spill with fire 43-Excessive Heat 53-Smoke odor removal 65-Steam, gas mistaken for smoke
54-Animal Rescue 71-Malicious false

13

ACTION TAKEN
1-Extinguishment 4-Remove Hazard 8-Fill in, Move up
2-Rescue or Assistance 5-Stand by 9-Not classified
3-Investigation only 6-Salvage 0-Undetermined
7-Ambulance

1

MUTUAL AID
1-Rec'd
2-Given
N/A

NA

FIXED PROPERTY USE (Occupancy) Pg 23-43   Trailer   10 19
IGNITION FACTOR Pg 44-45   Exposure Fire   65

CORRECT ADDRESS (Up to maximum of 21 characters) 3671 Dayton Park Dr
ZIP CODE 45414   **CENSUS TRACT** 80 7

OCCUPANT NAME (LAST, FIRST, MI)   Unknown   **TELEPHONE**   **ROOM or APT.**

OWNER NAME (LAST, FIRST, MI)   Unknown   **ADDRESS**   **TELEPHONE**

METHOD OF ALARM
1-Telephone direct 4-Radio 8-Voice signal municipal alarm signal
2-Municipal alarm system 5-Verbal 9-Not classified above
3-Private alarm system 6-No alarm recd 0-Undetermined or not reported
7-Tie-line (911)

3

CO. INSPECTION DISTRICT 12
SHIFT 2
NO. ALARMS 3

NO. FIRE SERVICE PERSONNEL RESPONDED 88
NO. ENGINES RESPONDED 10
NO. AERIAL APPARATUS RESPONDED 5
NO. OTHER VEHICLES RESPONDED 2

NUMBER OF INJURIES
FIRE SERVICE   OTHER
NUMBER OF FATALITIES
FIRE SERVICE   OTHER

COMPLEX Pg 61-62 No Complex 98
MOBILE PROPERTY TYPE Pg 63-65 (Complete Line S) NA = 08 17 Tractor Trailers 23

AREA OF FIRE ORIGIN Pg 67-68 Vehicle 47
EQUIPMENT INVOLVED IN IGNITION Pg 71-72 (Complete Line T) 98 No Equipment 98

FORM OF HEAT IGNITION Pg 74-76 Exposure 81
TYPE OF MATERIAL IGNITED Pg 78-79 Flammable Liquid 21
FORM OF MATERIAL IGNITED Pg 80-81 Bulk Storage 57

METHOD OF EXTINGUISHMENT
1-Self extinguished 5-Pre-connect hose/tank only
2-Make shift aids 6-Pre-connect hose/hydrant draft standpipe
3-Portable extinguisher 7-Hand-laid hose/hydrant draft standpipe
4-Automatic ext system 8-Master stream device
9-Not classified above
0-Undetermined or not reported

8

LEVEL OF FIRE ORIGIN
1-Grade level to 9 ft 6-Over 70 feet
2-10 to 19 feet 7-Objects in flight
3-20 to 29 feet 8-Below ground level
4-30 to 49 feet 9-Not classified above
5-50 to 70 feet 0-Undetermined

1

ESTIMATED TOTAL DOLLAR LOSS Undetermined 00 0

Number of Stories
1-1 story 4-5 to 6 stories 7-25 to 49 stories
2-2 story 5-7 to 12 stories 8-50 stories or more
3-3 to 4 stories 6-13 to 24 stories 0-Number of stories undetermined or not reported

1

CONSTRUCTION TYPE
1-Fire resistive 4-Unprotected noncombustible 8-Unprotected wood frame
2-Heavy timber 5-Protected ordinary 9-Not classified above
3-Protected noncombustible 6-Unprotected ordinary 0-Undetermined or not reported
7-Protected wood frame

0

EXTENT OF DAMAGE   Flame-Smoke
Confined to the object of origin 1 1
Confined to part of room or area of origin 2 2
Confined to room of origin 3 3
Confined to the fire-rated comp of origin 4 4
Confined to floor of origin 5 5
Confined to structure of origin 6 6
Extended beyond structure of origin 7 7
No damage of this type (N/A) 9 9
Undetermined or not reported 0 0

FLAME 7 P
SMOKE 7

DETECTOR PERFORMANCE
1-Det in room or space of fire origin - oper
2-Det not in rm or space of fire origin - oper
3-Det in rm or space of origin - no oper
4-Det not in rm or space of origin - not oper
5-Det in rm or space of fire origin but fire too small to oper
9-Not classified above
0-Undetermined or not reported
8-No detectors present (N/A)

8 - NA

SPRINKLER PERFORMANCE
1-Equipment operated
2-Equipment should have operated - did not
3-Equipment pre. but fire too small to oper.
9-Not classified above
0-Undetermined or not reported
8-No equipment present (N/A)

8 NA

TYPE OF MATERIAL GENERATING MOST SMOKE Pg 103-104
IF SMOKE SPREAD BEYOND ROOM OF ORIGIN   Flammable Liquid 21

AVENUE OF SMOKE TRAVEL
1-Air handling duct 4-Stairwell 7-Utility opening in floor
2-Corridor 5-Opening on construction 8-No avenue of smoke travel (N/A)
3-Elevator shaft 6-Utility opening in wall 0-Undetermined or not reported
9-Not classified above

5

FORM OF MATERIAL GENERATING MOST SMOKE Pg 108-109 Bulk Storage 57

IF MOBILE PROPERTY	YEAR	MAKE	MODEL	SERIAL NO.	LICENSE NO.
Tractor Trailers		Unknown		Unknown	

IF EQUIPMENT INVOLVED IN IGNITION	YEAR	MAKE	MODEL	SERIAL NO.	

MEMBER MAKING REPORT FF Michael Kenny   **DATE**
OFFICER IN CHARGE (if different) D.C. Don L. Douglas   **DATE** 5-27-87

Remarks _____

check if remarks continued on back   COM 6413

Appendix C (continued)

NFIRS-1

OHIO FIRE INCIDENT REPORTING SYSTEM

INCIDENT REPORT

Fire Department **Dayton Fire Dept**

1 ☐ DELETE
2 ☐ CHANGE

FDID	INCIDENT NO	EXP	MO	DAY	YEAR	DAY OF WEEK		ALARM TIME	ARRIVAL TIME	TIME IN SERVICE
5 7 0 1 1	2 5 2 4 0		0 5	2 7	8 7	1-Sunday 2-Monday 3-Tuesday / 4-Wednesday 5-Thursday 6-Friday / 7-Sat. **4**		2 1 0 7	2 1 1 1	0 7 0 0

GRASS, TRASH FIRES, SHORT FORM

SITUATION FOUND
11-Structure Fire
12-Outside Dollar Loss
13-Vehicle Fire
14-Brush, grass, leaves
15-Trash, Rubbish
16-Explosion, No after fire
17-Outside spill with fire

22-Air Gas Rupture
32-Emergency Medical call
33-Locked-in trapped
34-Search
35-Extrication
41-Spill, leak-No fire
42-Explosive, Bomb removal
43-Excessive Heat

44-Power line down
45-Arcing electric equipment
46-Aircraft standby
47-Chemical spill
51-Lock-out
52-Water removal
53-Smoke odor removal
54-Animal Rescue

55-Assist Police
56-Unauthorized burning
57-Move-up
59-Other service calls
61-Smoke scare
63-Controlled burn
65-Steam, gas mistaken for smoke
71-Malicious false

72-Bomb Scare
73-Alarm Malfunction
74-Unintentional false
99-Unclassified
Other _____ See page 17-19

13

ACTION TAKEN
1-Extinguishment
2-Rescue or Assistance
3-Investigation only
4-Remove Hazard
5-Stand by
6-Salvage
7-Ambulance
8-Fill in, Move up
9-Not classified
0-Undetermined

1

MUTUAL AID
1-Rec'd
2-Given
N/A

NA

FIXED PROPERTY USE (Occupancy) Pg 23-43
Parking lot 9 6 5

IGNITION FACTOR Pg 44-45
Exposure Fire 6 5

CORRECT ADDRESS (Up to maximum of 21 characters)
3671 Dayton Park Dr

ZIP CODE **4 5 4 1 4** | CENSUS TRACT **8 0 7**

COMPLETE FOR ALL INCIDENTS

OCCUPANT NAME (LAST, FIRST, MI)
Unknown
TELEPHONE | ROOM or APT.

OWNER NAME (LAST, FIRST, MI)
Unknown
ADDRESS | TELEPHONE

METHOD OF ALARM
1-Telephone direct
2-Municipal alarm system
3-Private alarm system
4-Radio
5-Verbal
6-No alarm recd.
7-Tie-line (911)
8-Voice signal municipal alarm signal
9-Not classified above
0-Undetermined or not reported

3

CO. INSPECTION DISTRICT | SHIFT | NO. ALARMS
1 2 | 2 | 3

NO. FIRE SERVICE PERSONNEL RESPONDED 8 4
NO. ENGINES RESPONDED 1 0
NO. AERIAL APPARATUS RESPONDED 8
NO. OTHER VEHICLES RESPONDED 1 2

COMPLETE IF CASUALTY

NUMBER OF INJURIES
FIRE SERVICE | OTHER
NUMBER OF FATALITIES
FIRE SERVICE | OTHER

COMPLEX Pg 61-62
No complex 1 8

MOBILE PROPERTY TYPE Pg 63-65 (Complete Line S) NA = 08
(3) Tractors 2 3

AREA OF FIRE ORIGIN Pg 67-68
Vehicle 4 7

EQUIPMENT INVOLVED IN IGNITION Pg 71-72 (Complete Line T) 98
No equipment 1 8

COMPLETE FOR ALL FIRES

FORM OF HEAT IGNITION Pg 74-76
Exposure 8 1

TYPE OF MATERIAL IGNITED Pg 78-79
Flammable Liquid 2 1

FORM OF MATERIAL IGNITED Pg 80-81
Bulk Storage 8 2

METHOD OF EXTINGUISHMENT
1-Self extinguished
2-Make shift aids
3-Portable extinguisher
4-Automatic ext. system
5-Pre-connect hose/tank only
6-Pre-connect hose/hydrant draft standpipe
7-Hand-laid hose/hydrant draft standpipe
8-Master stream device
9-Not classified above
0-Undetermined or not reported

8

LEVEL OF FIRE ORIGIN
1-Grade level to 9 ft.
2-10 to 19 feet
3-20 to 29 feet
4-30 to 49 feet
5-50 to 70 feet
6-Over 70 feet
7-Objects in flight
8-Below ground level
9-Not classified above
0-Undetermined

1

ESTIMATED TOTAL DOLLAR LOSS
undetermined 6 0 4 4

Number of Stories
1-1 story
2-2 story
3-3 to 4 stories
4-5 to 6 stories
5-7 to 12 stories
6-13 to 24 stories
7-25 to 49 stories
8-50 stories or more
0-Number of stories undetermined or not reported

1

CONSTRUCTION TYPE
1-Fire resistive
2-Heavy timber
3-Protected noncombustible
4-Unprotected noncombustible
5-Protected ordinary
6-Unprotected ordinary
7-Protected wood frame
8-Unprotected wood frame
9-Not classified above
0-Undetermined or not reported

0

COMPLETE IF STRUCTURE FIRE

EXTENT OF DAMAGE

	Flame	Smoke
Confined to the object of origin	1	1
Confined to part of room or area of origin	2	2
Confined to room of origin	3	3
Confined to fire-rated comp of origin	4	4
Confined to floor of origin	5	5
Confined to structure of origin	6	6
Extended beyond structure of origin	7	7
No damage of this type (N/A)	9	9
Undetermined or not reported	0	0

FLAME **1** P
SMOKE **7**

DETECTOR PERFORMANCE
1-Det. in room or space of fire origin - oper
2-Det. not in rm or space of fire origin - oper
3-Det. in rm or space of origin - no oper
4-Det. not in rm or space of origin - not oper
5-Det. in rm or space of fire origin but fire too small to oper.
9-Not classified above
0-Undetermined or not reported
8-No detectors present (N/A)

8

SPRINKLER PERFORMANCE
1-Equipment operated
2-Equipment should have operated - did not
3-Equipment pres. but fire too small to oper.
9-Not classified above
0-Undetermined or not reported
8-No equipment present (N/A)

8

TYPE OF MATERIAL GENERATING MOST SMOKE Pg 103-104
IF SMOKE SPREAD BEYOND ROOM OF ORIGIN
Flammable Liquid 2 1

AVENUE OF SMOKE TRAVEL
1-Air handling duct
2-Corridor
3-Elevator shaft
4-Stairwell
5-Opening on construction
6-Utility opening in wall
7-Utility opening in floor
8-No avenue of smoke travel (N/A)
0-Undetermined or not reported
9-Not classified above

0

FORM OF MATERIAL GENERATING MOST SMOKE Pg 108-109
Storage 5 2

	YEAR	MAKE	MODEL	SERIAL NO.	LICENSE NO.
IF MOBILE PROPERTY		(3)	International Tractor		
IF EQUIPMENT INVOLVED IN IGNITION					

MEMBER MAKING REPORT
FF Michael Kenny
DATE
OFFICER IN CHARGE (if different)
D.c. Dan S. Doylor
DATE 5-28-87

Remarks 1) International Oh Lic : 16 N 37 2) International
ALA Lic : X 8 65121 3) International ALA ALX 865182

☐ check if remarks continued on back

COM 9013

Appendix C (continued)

FIRE SERVICE CASUALTY REPORT

OHIO FIRE INCIDENT REPORTING SYSTEM   Fire Department _Dayton Fire Dept_   NFIR-3

1 ☐ DELETE REPORT
2 ☐ CHANGE

FDID	INCIDENT NO.	EXPOSURE NO.	CASUALTY NO.	INJURY OCCURRED	MO. DAY YEAR	TIME OF INJURY
5701/1	25304	100	12		5 28 87	1001 17

CASUALTY NAME (LAST, FIRST, MI) _Bergman   Mark_

TYPE OF CASUALTY
1. Fireground injury before F.D. arrival
2. Fireground injury after F.D. arrival
3. Injury during response to or return from incid.
4. Non-fire incident injury creating the alarm
5. Non-fire incident injury after alarm
6. Medical aid call, illness creating alarm
7. Not classified above
0. Undetermined or not reported

AGE _13_

SEX
1. ☒ Male
2. ☐ Female

CASE SEVERITY
1. Minor - The patient is not in danger of death or permanent disability. Immediate medical care is not necessary.
2. Moderate - There is little danger of death or permanent disability. Quick medical care is available. This category includes injuries such as fractures or lacerations requiring sutures.
3. Severe - The situation is potentially life threatening if the condition remains uncontrolled. Immediate medical care is necessary even though body processes may still be functioning and vital signs may be normal.
4. Life Threat - Death is imminent. Body processes and vital signs are not normal. Immediate medical care is necessary. This category includes such as severe hemorrhaging, multiple trauma and multiple internal injuries.
5. D.O.A. - Dead upon arrival at the scene   6. Died subsequent to arrival.

PRIMARY APPARENT SYMPTOM Pg. 151   _1_   STRAIN   51

PRIMARY PART OF BODY Pg. 153   _Leg_   41

PATIENT TAKEN TO
1. Hospital, emergency room or general admission
2. Doctor's office clinic
3. Long-term care facility
4. Morgue
5. Funeral home
6. Residence
7. Not transported
9. Not classified above
0. Undetermined or not reported

ASSIGNMENT
1. Fire suppression
2. Emergency Medical Service (EMS)
3. Fire prevention/inspection
4. Training
5. Maintenance
6. Fire alarm/communications
7. Administrative
0. Undetermined or not reported   _1_

NUMBER RESPONSES PRIOR TO INJURY
1. One
2. Two
3. Three
4. Four
5. Five
6. Six to eight
7. None to twelve
8. Over twelve
9. None
0. Undetermined or not reported   _3_

PHYSICAL CONDITION
1. Rested
2. Fatigued
3. Impaired (drugs, alcohol)
4. Impaired (illness, medication)
9. Not classified above
0. Undetermined or not reported   _2_

STATUS BEFORE ALARM
1. Asleep
2. Awake
9. Not classified above
0. Undetermined or not reported   _2_

FIRE FIGHTER ACTIVITY - Pg. 161   _Hose Lines_   31

WHERE INJURY OCCURRED - Pg. 163   _Outside   Ground Level_   2V

CAUSE OF FIRE FIGHTER INJURY - Pg. 165   _Lifting Hose_   501

MEDICAL CARE PROVIDED
1. None
2. Treated at scene
3. Treated at medical clinic
4. Treated at doctor's office
5. Hospital emergency room
6. Hospital outpatient
7. Hospital inpatient
8. Continued care after hospital release
9. Not classified above
0. Undetermined or not reported   _5_

PROTECTIVE COAT WORN
1. Nomex protective coat with liner
2. Nomex protective coat without liner
3. Canvas protective coat with liner
4. Canvas protective coat without liner
5. Rubber (or rubberized) coat with liner
6. Rubber (or rubberized) coat without liner
7. Other protective coat with liner
8. Other protective coat without liner
9. No protective coat being worn when injured
0. Undetermined or not reported   _1_

STATUS OF PROTECTIVE COAT
1. Open
2. Partially open
3. Closed, collar up
4. Closed, collar down
8. No protective coat being worn
9. Not classified above
0. Undetermined or not reported   _4_

PROBLEM WITH PROTECTIVE COAT
1. Burned
2. Ripped
3. Melted
4. Cut
5. Trapped steam or hazardous gases
7. No failure of the protective coat
8. No protective coat worn
9. Not classified above
0. Undetermined or not reported   _7_

PROTECTIVE TROUSERS WORN
1. Nomex protective trousers with liner
2. Nomex protective trousers without liner
3. Canvas protective trousers with liner
4. Canvas protective trousers without liner
5. Rubber (or rubberized) protective trousers with liner
6. Rubber (or rubberized) without liner
7. Other protective trousers with liner
8. Other protective trousers without liner
9. No protective trousers being worn
0. Undetermined or not reported   _1_

STATUS OF PROTECTIVE TROUSERS
1. Protective trousers worn properly
2. Protective trousers worn inside boots
3. Protective trousers worn without suspension
8. No protective trousers
9. Not classified above
0. Undetermined or not reported   _1_

PROBLEM WITH PROTECTIVE TROUSERS
1. Burned
2. Ripped
3. Melted
4. Cut
5. Trapped steam or hazardous gases
7. No failure of the protective trousers
8. No protective trousers worn
9. Not classified above
0. Undetermined or not reported   _7_

BOOTS/SHOES WORN
1. Boots, knee length (steel baseplate and steel toe)
2. Boots, knee length (steel toe only)
3. Boots, ¾ length (steel baseplate and steel toe)
4. boots, ¾ length (steel toe only)
5. Shoes, safety (steel baseplate and steel toe)
6. Shoes, safety (steel toe only)
7. Shoes, without steel reinforcement
8. Shoes, non-safety
9. Not classified above
0. Not classified or not reported   _1_

STATUS OF BOOTS
1. ¾ length boots pulled up (full length)
2. ¾ length boots not pulled up
3. Knee length boots worn
8. No boots worn
9. Not classified above
0. Undetermined or not reported   _3_

PROBLEM WITH BOOTS/SHOES
1. Burned
2. Ripped
3. Cut
4. Punctured
5. Object fell into
6. Failed under impact
8. No failure of boots/shoes
9. No boots/shoes
0. Undetermined or not reported   _8_

HELMET WORN
1. Leather helmet
2. Aluminum helmet
3. Glass fiber helmet
4. Polycarbonate helmet
8. No helmet being worn
9. Not classified above
0. Undetermined or not reported   _4_

STATUS OF HELMET
1. Chin strap in use
2. Chin strap and ear/neck protector in use
3. Ear/neck protector only in use
4. Chin strap and ear/neck protector not in use
8. No helmet being worn
9. Not classified above
0. Undetermined or not reported   _1_

PROBLEM WITH HELMET
1. Burned
2. Melted
3. Fractured
4. Punctured
5. Knocked off
7. No failure of helmet
8. No helmet worn
9. Not classified above
0. Undetermined or not reported   _7_

FACE PROTECTION WORN
1. Full face protection
2. Partial face protection
3. Goggles worn
8. No face protection worn
9. Not classified above
0. Undetermined or not reported   _8_

PROBLEM WITH FACE PROTECTION
1. Burned
2. Melted
3. Fractured/cracked/broke
4. Scratched
7. No failure of face protection
8. No face protection being used
9. Not classified above
0. Undetermined or not reported   _7_

BREATHING APPARATUS WORN
1. Self-contained open circuit demand type apparatus
2. Self-contained open circuit pressure type apparatus
3. Self-contained closed circuit type apparatus
4. Not self-contained
8. No breathing apparatus being used
9. Not classified above
0. Undetermined or not reported   _8_

STATUS OF BREATHING APPARATUS
1. Face piece and regulator connected
2. Air supply turned off
3. Face piece not in place
4. Harness not secured
5. Breathing apparatus properly worn
8. No breathing apparatus
9. Not classified above
0. Undetermined or not reported   _8_

PROBLEM WITH BREATHING APPARATUS   _None_   918

GLOVES WORN
3. Canvas
6. Rubber
1. Cotton
4. Leather
7. Synthetic, including nomex
2. Wool
5. Asbestos
8. No gloves being worn
9. Not classified above
0. Undetermined or not reported   _4_

PROBLEM WITH GLOVES
1. Burned
2. Ripped
3. Melted
4. Cut/punctured
5. Object fell into
6. Insufficient insulation
7. No failure of the gloves
8. No gloves being worn
9. Not classified above
0. Undetermined or not reported   _7_

SPECIAL EQUIPMENT WORN
1. Proximity suit
2. Chemical suit
3. Scuba gear
4. Exposure suit
5. Life preservers
6. Life belt, ladder belt
7. Personnel lighting
8. No special equipment being used
9. Not classified above
0. Undetermined or not reported   _8_

STATUS OF SPECIAL EQUIPMENT
1. Being worn properly and used for designed purpose
2. Being worn properly but not being used for designed purpose
3. Not being worn properly but used for designed purpose
4. Not being worn properly and not used for designed purpose
8. No special equipment being used   _8_

PROBLEM WITH SPECIAL EQUIPMENT
1. Burned
2. Ripped, torn, cut, punctured
3. Melted
4. Not properly serviced/stored prior to use
5. Not used for designed purpose
6. Not used as recommended by manufacturer
7. No problem with special equipment
8. No special equipment being used
9. Not classified above
0. Undetermined or not reported   _8_

OFFICER IN CHARGE	DATE	MEMBER MAKING REPORT	DATE
D.C. Gary L. Dowler	5-28-87	FF. Michael R Kenny	5/28/87

Remarks ——

☐ CHECK IF REMARKS CONTINUED ON BACK   COM 5013

Appendix C (continued)

OHIO FIRE INCIDENT REPORTING SYSTEM

CIVILIAN CASUALTY REPORT

NFIRS-2

Fill In This Report In Your Own Words

Fire Department __Dayton Fire Dept__

FDID	Incident No.	Exp. No.	Mo.	Day	Year	Day of the Week	Alarm Time
57011	175244	00	05	27	87	Wed	18 21/07

CASUALTY SEVERE ENOUGH TO CHECK ON LATER YES ☑ NO ☐

ENTER CORRECT CODE NUMBER IN BOX

CASUALTY NUMBER 11

1 ☐ DELETE 2 ☐ CHANGE

CASUALTY LAST NAME	FIRST NAME	MI	D.O.B.	AGE	TIME OF INJURY
Griffith	Curtis		Unknown	29	21/010

HOME ADDRESS 51 2720 Sherer

TELEPHONE 274-5649

SEX
1 Male
2 Female

[1]

CASUALTY TYPE
1 Fire Casualty
2 Action Casualty
3 EMS Casualty

[1]

SEVERITY
1 Injury
2 Death

[1]

AFFILIATION
1 Fire Service
2 Other Emergency Personnel
3 Civilian

[3]

FAMILIARITY WITH STRUCTURE
1 Less than 1 Day
2 1 to 7 Days
3 8 to 30 Days   0 Undetermined or
4 1 to 2 Months   not reported
5 3 to 6 Months
6 7 to 12 Months
7 Over 1 Year
8 Not a Structure

[0]

LOCATION AT IGNITION
1 Intimately involved with ignition
2 In the room or space of fire
3 On same floor as origin of fire
4 In same building as origin of fire
5 Outside of building of fire origin but on property
6 Fire casualty off property of fire origin
0 Undetermined or not reported
8 Not a fire casualty
9 Not classified above

[1]

CONDITION BEFORE INJURY
1 Asleep
2 Bedridden, other physical handicap
3 Impaired by drugs, alcohol
4 Under restraint
5 Too young to act
6 Too old to act
7 Mentally handicapped, senile
0 Undetermined or not reported
8 Awake, unimpaired
9 Not classified above

[8]

CONDITION PREVENTING ESCAPE
1 No time to escape, explosion or fire progressed too rapidly
2 Fire between casualty and exit
3 Locked doors
4 Illegal gates, locks
5 Clothing and casualty burning
6 Moved too slowly
7 Victim incapacitated prior to ignition
8 No conditions prevented escape or not a factor
9 Not classified above
0 Undetermined or not reported

[1]

ACTIVITY AT TIME OF INJURY
1 Escaping
2 Rescue attempt
3 Fire control
4 Response/return
5 Cleanup, salvage, mop-up
6 Sleeping
7 Unable to act
0 Undetermined or not reported
8 Irrational action
9 Not classified above

[5]

CAUSE OF INJURY
1 Caught in, under, between trapped by
2 Exposed to fire products
3 Exposed to chemical radiation
4 Fell or stepped on, over, into
5 Overexertion
6 Rubbed by, contact with
7 Struck by
9 Not classified above
0 Undetermined or not reported
8 Not applicable

[2]

NATURE OF INJURY
1 Burns asphyxia/smoke
2 Burns only
3 Asphyxia smoke only
4 Wound, cut, bleeding
5 Dislocation, fracture
6 Complaint of pain
7 Shock
8 Strain, sprain
9 Not classified above
0 Undetermined or not reported

[2]

PART OF BODY INJURED
1 Head, neck
2 Body, trunk, back
3 Arm
4 Leg
5 Hand
6 Foot
7 Internal Included are respiratory system, heart
8 Multiple parts
9 Not classified above
0 Undetermined or not reported

[4]

DISPOSITION
1 Refused help
2 Treated at scene and released
3 Taken to hospital by fire dept. vehicle
4 Taken to hospital by non fire dept. vehicle
5 Taken to other than a hospital
6 Died
7 Not classified above
0 Undetermined or not reported

[3]

☐ SEE REMARKS ON BACK   ☐ SEE ADDITIONAL REPORT

CASUALTY SEVERE ENOUGH TO CHECK ON LATER YES ☐ NO ☐

ENTER CORRECT CODE NUMBER IN BOX

CASUALTY NUMBER

1 ☐ DELETE 2 ☐ CHANGE

CASUALTY LAST NAME	FIRST NAME	MI	D.O.B.	AGE	TIME OF INJURY

HOME ADDRESS 51

TELEPHONE

SEX
1 Male
2 Female

CASUALTY TYPE
1 Fire Casualty
2 Action Casualty
3 EMS Casualty

SEVERITY
1 Injury
2 Death

AFFILIATION
1 Fire Service
2 Other Emergency Personnel
3 Civilian

FAMILIARITY WITH STRUCTURE
1 Less than 1 Day
2 1 to 7 Days
3 8 to 30 Days   0 Undetermined or
4 1 to 2 Months   not reported
5 3 to 6 Months
6 7 to 12 Months
7 Over 1 Year
8 Not a Structure

LOCATION AT IGNITION
1 Intimately involved with ignition
2 In the room or space of fire
3 On same floor as origin of fire
4 In same building as origin of fire
5 Outside of building of fire origin but on property
6 Fire casualty off property of fire origin
0 Undetermined or not reported
8 Not a fire casualty
9 Not classified above

CONDITION BEFORE INJURY
1 Asleep
2 Bedridden, other physical handicap
3 Impaired by drugs, alcohol
4 Under restraint
5 Too young to act
6 Too old to act
7 Mentally handicapped, senile
0 Undetermined or not reported
8 Awake, unimpaired
9 Not classified above

CONDITION PREVENTING ESCAPE
1 No time to escape; explosion or fire progressed too rapidly
2 Fire between casualty and exit
3 Locked doors
4 Illegal gates, locks
5 Clothing and casualty burning
6 Moved too slowly
7 Victim incapacitated prior to ignition
8 No conditions prevented escape or not a factor
9 Not classified above
0 Undetermined or not reported

ACTIVITY AT TIME OF INJURY
1 Escaping
2 Rescue attempt
3 Fire control
4 Response/return
5 Cleanup, salvage, mop-up
6 Sleeping
7 Unable to act
0 Undetermined or not reported
8 Irrational action
9 Not classified above

CAUSE OF INJURY
1 Caught in, under, between trapped by
2 Exposed to fire products
3 Exposed to chemical radiation
4 Fell or stepped on, over, into
5 Overexertion
6 Rubbed by, contact with
7 Struck by
9 Not classified above
0 Undetermined or not reported
8 Not applicable

NATURE OF INJURY
1 Burns asphyxia/smoke
2 Burns only
3 Asphyxia smoke only
4 Wound, cut, bleeding
5 Dislocation, fracture
6 Complaint of pain
7 Shock
8 Strain, sprain
9 Not classified above
0 Undetermined or not reported

PART OF BODY INJURED
1 Head, neck
2 Body, trunk, back
3 Arm
4 Leg
5 Hand
6 Foot
7 Internal Included are respiratory system, heart
8 Multiple parts
9 Not classified above
0 Undetermined or not reported

DISPOSITION
1 Refused help
2 Treated at scene and released
3 Taken to hospital by fire dept. vehicle
4 Taken to hospital by non fire dept. vehicle
5 Taken to other than a hospital
6 Died
7 Not classified above
0 Undetermined or not reported

☐ SEE REMARKS ON BACK   ☐ SEE ADDITIONAL REPORT

OFFICER IN CHARGE AT INCIDENT (Name, Position)	Date	MEMBER MAKING REPORT	Date
D.C. Gary L. Dougler	5-27-87	FF Michael R Kenny	

COM 5013

Appendix C (continued)

Fire-building approximately 180,000 square feet, one-story tilt-up construction on slab. Steel bar joist roof on center steel post supports, approximately 30 feet high. Building used for warehousing auto paint finishes, with thinners, additives, etc. Exposure on east: approximately 1,400 drums of thinners-lacquers-paints stacked two-three high on pallets. One-story office area also on east side of warehouse.

ADT alarm on initial dispatch; also dispatcher reported several phone calls. Second alarm staged at Wagner Ford when large column of black smoke visible from Keowee Street. Upon arrival had third alarm staged.

On arrival, flame showing through roof and entire east half of building involved. Engine 12 on hydrant and supplied Engine 8 deck gun to protect office and drum storage on east side. Engine 21 on hydrant in front of structure; supplied Engine 14 and Truck 14 to protect trailers in dock and parking areas. Aerosal cans raining on crews. Decision not to supply sprinkler system: severe exposure hazard so close to building with sprinkler connections facing fire building--especially since sprinkler piping already probably compromised.

Truck 11 assigned to east sector, under command of Central 2. West 2 assigned as planning sector. Unmanned monitors set up on east exposures as fire progressed.

Tactical decision to attempt to stop fire at north-south firewall in center of structure. However, planning sector found heavy extension into west side before any actual company assignments could be made.

Occupants of structure, approximately 30, reported an employee injury to arriving firefighters, and Engine 4 assisted Medic 4 with lift--truck operator burned seriously on legs. Engine 4 later assigned as brand patrol.

At approximately 2150 Chief 1 took command, made Chief 3 operation sector. Chief 1's strategy was to maintain streams on exposures--which were all on drained concrete pad--but to throw no water on burning structure. The building sits directly over city wellfields, and possible contamination of water supply from run-off became first priority.

Warehouse area of structure a total loss. Drum storage never involved in fire, though some damage later as walls collapsed. Offices sustained some minor water damage, but records preserved and recovered. Seventeen truck trailers on scene heavily or totally damaged; dozens of others not damaged and later removed from parking area.

Injury to firefighter minor (pulled muscle).

Fire officially contained at 0012 on 5-28-87, and under control at 1004 on 6-2-87.

REMARKS:

Fire allowed to burn freely until self-extinguished: Goal to not contaminate water supply:

Thursday (28th): Met with Sherwin-Williams and OEPA, RAPCA, other interested parties, to determine effect of fire on environment, while developing plan to continue protection of exposures on east side of building. Fire still free burning. Flare up caused staging of a 2nd alarm assignment at site. Run-off to Miami River contained and solvents skimmed and pumped off. Basic operation still

Appendix C (continued)

geared around protecting well field. Command trailer installed. Valuables in office area turned over to Sherwin-Williams representatives.

Friday (29th): Met with federal and state EPA and RAPCA to secure site safety plan. Smoke and fire diminishing, but still visible throughout complex. Sherwin-Williams given permission to remove undamaged semis and trailers.

Saturday (30th): Sherwin-Williams contracted with O-H Material to be prime clean up contractor. Environmental monitoring showed no contamination of groundwater, but some of ground in site and river near drains. Sherwin-Williams given permission to remove all items in the office area. Dave Tabar of Sherwin-Williams removed inventory sheet from command post. Total inventory as of 5-16-87 in excess of 1.7 million gallons.

Sunday (31st): Removal of trailers next to building. Walls dismantled. Fire down to 5-6 spots, glow with light smoke. Site safety plan supposed to be developed by O-H Material for review by fire department. Containment dike for runoff under construction.

Monday (1st): Containment dike completed. Fire still smolders--some hot spots and vapors. Some barrels removed from east side--removal of trailers. Site safety plan in place. Clean up to start at No. 1 door at front loading dock. Work during daylight only. Reduced fire department to one engine and one truck at night. Barrel removal halted until proper forklift is at site.

Tuesday (2nd): Fire placed under control and loss established.

APPENDIX E

TELEPHONE BOARD #1

Dispatcher:	2105 Yeah, Fire Box No. 455 goes to Sherwin-Williams, 3671 Dayton Park Dr.
Dispatcher:	What was that, 30 what
Caller:	3671 Dayton Park Road
Dispatcher:	455
Caller:	Right
Dispatcher:	OK, are you going to have a runner enroute Got one enroute now
Dispatcher:	OK, we'll get them on the way
Dispatcher:	2106 Dayton Fire Department
Caller:	I want to report a fire at the Sherwin-Williams Warehouse on Dayton Park Drive
Dispatcher:	We got them on the way sir
Caller:	OK, you might want to send some ambulances
Dispatcher:	2107 Dayton Fire Department
Caller:	Yes, I'm an employee of B & O Railroad, there's a fire at Dayton Industrial Park, has there been a report
Dispatcher:	At Sherwin-Williams
Caller:	Ah, I guess I'm not sure where it's at
Dispatcher:	There on Dayton Park Road, we got them on the way sir
Caller:	All right, thank you
Dispatcher:	Thank you
Dispatcher:	2108 Dayton Fire
Caller:	Yeah, there's a fire out here, I'm at Troy and Stanley.
Dispatcher:	Yes ma'am we've got them on the way.
Caller:	Thank you.
Dispatcher:	2108 Dayton Fire
Caller:	Yes I have an emergency, need ah fire, we have a fire at 3671 Dayton Park Drive (employee)

25

Appendix E (continued)

Dispatcher:	They're on their way ma'am
Caller:	Thank you very much.
Dispatcher:	Bye.
Dispatcher:	Yeah, Dispatch
Caller:	There is a fire, I think it is right in front of North Lake Hills but I can't tell from where I'm standing
Dispatcher:	Over by Chuck Wagon Lane, over in that area
Caller:	North Lake Hills, Old Troy Pike
Dispatcher:	Yeah, we've got them on the way ma'am
Caller:	OK, thank you
Dispatcher:	OK FD Dispatcher calling Police Dispatcher
Dispatch	(Police)
FD Dispatcher:	We need a little assistance out at Wagoner Ford and Needmore
PD Dispatcher:	Uh huh, do you need any traffic control there today
FD Dispatcher:	Ah, probably going to
PD Dispatcher:	OK.
FD Dispatcher:	It's on Dayton Park Road is where the fire is Dayton Park Road. OK
PD Dispatcher:	OK, thank you
FD Dispatcher:	Dispatch
Company #2	Ah, Rescue 1 is back in quarters, at Co. 2's and I've got four guys if you want me to man an Engine 2, or I've got four guys however you want me to do, the truck crew my engine crew went ahead and put the truck in service and took that
FD Dispatcher:	OK, you might as well put the engine in service then
Company #2	OK, we'll be in service with Engine 2
FD Dispatcher:	OK, thanks.
Dispatcher:	2117 Dayton Fire
Caller:	Yeah, this is Mr.___speaking, have you got a report of a fire off Troy Street
Dispatcher:	Yes we do sir, we've got them on the way
Caller:	OK, that's really smoking

Appendix E (continued)

Dispatcher:	2118 Dayton Fire
Caller:	Yes, I'm sure you're aware there's a fire at Sherwin-Williams on Dayton Park Drive
Dispatcher:	Yes sir, we are.
Caller:	OK, now we're directly across the street from there, is there any danger to my people working there
Dispatcher:	As far as I know, at this time sir, I can't say, but if there is apparently any danger, I'm sure that they will evacuate
Caller:	OK, I can leave my people working
Dispatcher:	As far as I know
Caller:	OK
Dispatcher:	We've got people out there working, you know, if they see if things are dangerous, they'll get them out
Caller:	OK
Dispatcher:	2118 Dayton Fire
Caller:	Yeah, could you tell me are paramedics or ambulance on the way to 3671 Dayton Park Drive
Dispatcher:	Yes ma'am
Caller:	OK, cause we got a guy burning, and I didn't know if the ambulance had -
Dispatcher:	They're on the way
Caller:	OK, thanks
Dispatcher:	2119 Dayton Fire
Caller:	Yes sir, we're up on Earnst and North Main is there a fire burning on the other side of Riverside somewhere
Dispatcher:	Yes sir, there is, and I'm too busy to talk to you at this time sir
Caller:	OK, thank you
Dispatcher:	Yes sir.
Caller:	Has anybody reported a fire out here on Brandt Pike
Dispatcher:	Yes
Caller:	OK, thanks
Dispatcher:	2120 Dayton Fire
Caller:	I don't know if anybody's called or not, I live at 2329 Troy St.

Appendix E (continued)

Dispatcher:	Yeah.
Caller:	And, ah behind the trailers back here, there's something burning back here
Dispatcher:	Yes sir, we're aware of it, they're on the scene
Caller:	OK, thank you
Dispatcher:	2121 Dayton Fire
Caller:	Yes, I live at 104 Delaware looking out my back door, toward Main Street, there's a huge _____ of black smoke and
Dispatcher:	Yes ma'am, we're well aware of that, we have fire equipment on the scene there
Caller:	Oh, OK
Dispatcher:	2121 Dayton Fire
Caller:	Yes, we have a fire here off of Troy Street, have you gotten that
Dispatcher:	Yes, ma'am, they're there
Caller:	OK, then, I didn't hear the fire trucks, I thought I better call
Dispatcher:	Thank you ma'am
Caller:	Thank you.
Dispatcher:	2121 Dayton Fire
Caller:	Yes, I live out around 202 and I'm in an apartment building and see flames coming above the apartment building, I'm not quite sure how far over it is, but I know it's on Route 202
Dispatcher:	Yes ma'am, we've got fire crews on the scene over there
Caller:	Already?
Dispatcher:	Yes ma'am
Caller:	Thank you, bye bye
Dispatcher:	2122 Dayton Fire
Caller:	Has anybody reported a fire across the street from the _____
Dispatcher:	Yes ma'am, they have
Caller:	OK, thank you
Dispatcher:	2124 Dayton Fire
Caller:	Yes, I'm _____ on St. Adalbert across from the Sohio Oil thing, and it's on fire
Dispatcher:	There's a fire out there, yes ma'am, we've got equipment on the scene

Appendix E (continued)

Caller: OK, with me living this close, should I leave

Dispatcher: At this time ma'am, I would say no

Caller: OK, ah, will we be contacted if we should

Dispatcher: Yes ma'am, you will be

Caller: OK, thank you

Dispatcher: Yes ma'am, bye

Dispatcher: 2124 Dayton Fire

Caller: Yes, I live on Vermont Street, and I can see that flames and I was wondering

Dispatcher: Yes ma'am, we've got crews out there now

Caller: Well everybody on the street is taking off, and I was

Dispatcher: Well, I'm sorry ma'am, I really don't have time to talk to you, as far as I can tell at this time you're in no danger

Caller: Oh, can you tell me what it is

Dispatcher: It's a building out there burning, ma'am

Caller: Is it chemicals or

Dispatcher: Ma'am I don't know, I'm not there, I'm sorry I'm curt with you but I'm awfully busy right now

Caller: We're in no danger

Dispatcher: No ma'am

Caller: All right, thank you

Dispatcher: 2125 Dayton Fire

Caller: Ah, yes I want to report a possible fire in the, in the 200 block of Baltimore Street

Dispatcher: 200 block of where

Caller: Baltimore Street

Dispatcher: Beckmore? I'll I

Caller: B-A-L-T-I-M-O-R-E

Dispatcher: Oh, Baltimore

Caller: Right across from the park, Patterson Park

Dispatcher: Ah, do you have any idea what's burning

Appendix E (continued)

Caller: No, I don't know, I just see smoke up in the air and there's a bunch of flames shooting down there, I don't know what it is but I live in the 100 block and some of the neighbors next door went up that way, and I said well I'll go head and call

Dispatcher: Well, we have a fire out that way, we've got crews on the scene

Caller: Oh, do you, Oh, I didn't know, then I was just making sure someone knew about it

Dispatcher: OK, thank you

Caller: OK, thank you

Dispatcher: 2125 Dayton Fire

Caller: Hi, this is Joe _____ night supervisor at Earnst Enterprises on Wagoner Ford Road

Dispatcher: Yes, sir.

Caller: I was wondering if I should get my men out of here or not

I noticed that fire down there at AGA or whatever it's at

Dispatcher: Well, all I can tell you at this time sir, is they haven't said anything to us about any evacuations

Caller: OK

Dispatcher: I'm sure if it gets to the point that somebody should be evacuated they will do it

Caller: All right

Dispatcher: 2126 Dayton Fire

Caller: Hello, I was wondering if you have any information about a fire on Wagoner Ford Road

Dispatcher: No sir, I have no information at this time

Caller: You don't know what it is that's burning there

Dispatcher: No sir, and I don't have time to talk about it

Caller: OK

Dispatcher: Thank you

Dispatcher: 2126 Dayton Fire

Caller: Yeah, this is Greg from Preston Trucking, we're pretty close to that fire that is down the street, ah is that that chemical place

Dispatcher: Ah, Sherwin-Williams Paints

Appendix E (continued)

Caller:	Sherwin-Williams
Dispatcher:	Yeah
Caller:	Is there going to be any evacuation
Dispatcher:	Ah, at this time we haven't heard anything about any anticipation of it
Caller:	OK, cause we're pretty close
Dispatcher:	Well, I'm sure if they feel that it's necessary, they will evacuate the other people in the area sir
Caller:	Thank you very much
Dispatcher:	Yes Sir
Dispatcher:	2126 Dayton Fire
Caller:	Yes, are you aware of the fire at Sohio
Dispatcher:	It's not Sohio sir, and yes we have equipment out there
Caller:	Is it should we leave the area
Dispatcher:	Ah, at this time I would say no sir, they have not been any alarm for evacuation
Caller:	And it's not Sohio
Dispatcher:	No it's not
Caller:	OK
Dispatcher:	Sir, I'm sorry I don't have time to talk to you I'm really busy
Dispatcher:	2127 Dayton Fire
Caller:	OK, I live at 3801 South Shore Drive, in Dayton, and there's a fire over the apartment across from my window I didn't know if anyone had called
Dispatcher:	Are you sure it's coming from the apartment or are you looking over the top of it
Caller:	I'm looking over the top of the apartment across the yard from me
Dispatcher:	OK, and you're seeing an extreme amount of smoke and flames
Caller:	I'm seeing flames and smoke, yes black smoke
Dispatcher:	OK, could that fire you're looking at be over ah off Wagoner Ford Road
Caller:	I don't think so, I think it's right here in this housing development, cause it's the only thing I can see from where I'm at
Dispatcher:	3801 South Shore
Caller:	Yeah, that's my address, I don't know what that address is over there but if you come around here, you'll see it yourself

Appendix E (continued)

Dispatcher:	Well we've got a big fire down on Needmore Road there and the flames may be what you're looking at
Caller:	I don't know
Dispatcher:	What's your phone number
Caller:	My phone number here is 237-XXXX
Dispatcher:	OK, have you attempted to walk over to that building and see
Caller:	No my family lives there and I told them to call the Fire Department and they ran out of the house so I thought I'd call, I don't know, maybe I'm just being an alarmist maybe you're right, you know, but
Dispatcher:	38 across from 3801 South Shore, we'll get somebody to check it out
Caller:	OK, thank you
Dispatcher:	2129 Dayton Fire
Caller:	Yeah, this is Jerry over at Andy's, you got a report on this fire over here
Dispatcher:	Oh, yes sir
Caller:	Where's that at over here, I can see flames
Dispatcher:	Over off Needmore Road
Caller:	Off of Needmore, all right
Dispatcher:	2129 Dayton Fire
Caller:	Ah, hi, I live at 606 Brandt, which is right across the street from all the gas and oil tanks, has anybody reported a fire
Dispatcher:	They sure have ma'am
Caller:	Oh, OK, 'cause we haven't seen any action yet, OK, thanks Right
Dispatcher:	2129 Dayton Fire
Caller:	Do you have anything about a fire on Wagoner Ford Road
Dispatcher:	Yes sir we do
Caller:	Is there any point in anybody being evacuated at this time
Dispatcher:	Not at this time no sir
Caller:	Well, OK, we live close to it and I'm just wondering if we should be out or anything breathing it
Dispatcher:	No, I don't know of any reason for alarm at this time
Caller:	OK, thank you

Appendix E (continued)

Dispatcher: 2130 Dayton Fire

Caller: Yes sir, that fire's that's on the east end, can you tell me if that's some that's coming from that is toxic or not

Dispatcher: No

Caller: You don't know

Dispatcher: As far as I can tell you at this time, we have no reason for alarm

Caller: OK, I just wanted to check, I had some kids out playing and

Dispatcher: Right I can understand that sir

(Tape Transcription from Dispatch Log Tape Ended)—2130 Hours

Appendix E (continued)

TELEPHONE BOARD #2

Dispatcher: 2106 Dayton Fire Department

Caller: Ah, yeah I don't know if there's been a report but there's a fire at the Sherwin-Williams Warehouse at on Dayton Park Drive

Dispatcher: They're on the way sir

Caller: OK, ah you might want to send some ambulances down there too

Dispatcher: Where's that

Caller: The same place

Dispatcher: Why's that, are there a lot of people still there

Caller: Oh, yeah, they're working

Dispatcher: Do they know it's on fire

Caller: They're out of the place but it's burning bad

Dispatcher: OK

Caller: Thank you

Dispatcher: Thank you

Dispatcher: 2107 Dayton Fire Department

Caller: Yeah, this is Ted _____ at Kittyhawk Golf Course, there's a big explosion

Dispatcher: Yeah, we've got them on the way, Sherwin-Williams

Caller: Yeah, right off Wagoner Ford Road

Dispatcher: Yeah, we got them on the way

Caller: OK

Dispatcher: Thank you sir

Dispatcher: 2109 Dayton Fire Department

Caller: Ah, yes we need a fire truck out on Troy Pike, Troy Street you know where North Lake Hills is, right across the street from North Lake Hills, they have a great big old bundle of fire

Dispatcher: What's burning

Caller: I have no idea, my husband just went over there to find out

Dispatcher: Now we've got a fire over on Dayton Park Drive, is that what he's seeing

Caller: Dayton Park Drive, is that close to Troy Street

Appendix E (continued)

Dispatcher:	Well yeah, you can see it from there, it would be over there by the golf course, is it a building
Caller:	I can't tell we just see big black smoke coming up in the air
Dispatcher:	OK, that would be over there at that Dayton Park Drive, we've got a second alarm fire over there and from where you're at you can see straight through
Caller:	OK, I just wanted to make sure
Dispatcher:	OK, thank you

(Several calls from residences--interrupted on transcription by hearing Radio and PA conversation)

Dispatcher:	2112 Dayton Fire Department
Caller:	I know you're busy, Yes, this is Patty from Huber Heights Fire, do you have a fire in the area of Needmore and Wagoner Ford.
Dispatcher:	Yeah, up there on Chuck Wagon Lane
Caller:	Bye

Dispatcher calling Chief 3

Chiefs son:	Hello
Dispatcher:	Yeah, Paul, we've got, Paul
Chiefs son:	No, this is his son I don't know where he is right now
Dispatcher:	OK, if you can get a hold of him, tell him we have a third alarm fire at Dayton Park Drive
Chiefs son:	Dayton Park, all right
Dispatch:	We've got a third alarm at Sherwin-Williams over on Dayton Park Drive, I'm on my way

Dispatch calling Chief 1

Chief's Wife:	He's on the other phone, he'll be with you in a minute
Chief 1:	Send 15's on up will you please
Dispatcher:	Chief
Chief 1:	Just send 15's on up I'm hearing it

Appendix E (continued)

Dispatcher:	OK
Chief 1:	Thank you
Dispatcher:	2116
Caller:	Ah, yes sir, anybody called in for that fire over
Dispatcher:	Yes they have, they're already over there
Caller:	All right, thank you
Dispatcher:	2119 Dayton Fire Department
Caller:	Yeah, there's a fire over on, I think it's Commerce Park
Dispatcher:	Yeah, we've got a third alarm fire going ma'am
Caller:	Better hurry
Dispatcher:	They're there, they're there
Dispatcher:	2119 Dayton Fire Department
Caller:	I'm calling from 150 Jenny Road
Dispatcher:	Yeah, we've got crews on the scene for a big fire over there
Caller:	OK, I called to make sure
Chief 1:	What companies do you have
Dispatcher:	Hold on just a second
Dispatcher:	Coleman Yeah, Chief
Chief 1:	What companies do you have in reserve
Dispatcher:	Ah, the way it looks right now we don't have anybody left
Chief 1:	No, no what reserve apparatus, what companies have reserve apparatus

(Call transferred to Supervisor's position in Dispatch Center - not recorded on Dispatch Log Tape)

Dispatcher:	2119 Dayton Fire Department
Caller:	Ah, yes I live off of Valley Street, looking from Valley over toward Brandt you have a big fire
Dispatcher:	Yeah, we've got a third alarm fire going sir
Caller:	Oh, sorry
Dispatcher:	2124 Dayton Fire Department
Caller:	Yes, have you been called about the fire on _____
Dispatcher:	Yes we have

Appendix E (continued)

Caller:	OK, thank you
Dispatcher:	2124 Dayton Fire
Caller:	Yeah, I live on Vermont Street and I can see the flames and I wondered
Dispatcher:	Hung up
Dispatcher:	2124 Dayton Fire
Caller:	Yeah, we're at R & R over on Valley have you got a report of a fire
Dispatcher:	Yeah, it's over on it's by the Golf Course
Caller:	Golf Course
Dispatcher:	Dispatch
Co. 15:	Chief 1 just come to 15's and told us to call and tell you that Engine 15 is sitting here in the barn
Dispatcher:	Well they shouldn't be
Co. 15:	Well it is
Dispatcher:	Huh
Co. 15:	They told him he called down there they said they you guys told somebody that we weren't that there was nobody here we have not been dispatched, we're still here
Dispatcher:	OK, thanks

(Tape Transcription from Dispatch Log Tape Ended)--2126 Hours

RADIO AND PA

Dispatcher:	Box 455 that will be at Sherwin-Williams Paints, 3671 Dayton Park Road
	That's Box 455, that will be Sherwin-Williams Paints, 3671 Dayton Park Road, we've received a couple calls on this, Chuck Wagon Lane will be your cross
	That will be Engines 12, 21, 8, Truck 14 and the East Chief Engines 12, 21, 8 14, Truck 14, and the East Chief OK, Engine 12, 21, 8, Truck 14, East Chief All Clear 2107 Bowersock

Appendix E (continued)

Dispatcher:	Dispatch to East 2 we're still receiving numerous calls, says there's an explosion and quite a bit of fire at Sherwin-Williams
East 3:	Clearly, there's a large amount of smoke in the area, go head and dispatch me a full second alarm, stage them to the entrance there by the Kittyhawk Golf Course
Dispatcher:	2108
East 2:	East 2 Dispatcher, go head and start me that second chief, make sure you notify Chief 3
Dispatcher:	2108
Central 2:	Central 2 Dispatcher, Central 2's responding
Dispatcher:	OK, Central 2, 2109
Dispatcher:	All companies, we're on master, we have a fire at Sherwin-Williams Paints, 3671 Dayton Park Road, first alarm response is enroute we're dispatching Engine 14, 4, 2, Engine 18, Truck 2, Truck 11, Central Chief
Dispatcher:	That's Engines 14, 4, 2, 18, Truck 2, Truck 11 and the Central Chief OK, Truck 2, Truck 11 you clear on the air
Dispatcher:	Dispatcher, Engine 14 is clear and responding OK Engine 14, 4, 2, 18's, Central Chief clear, Truck 11 are you clear on the air? 2110 Bowersock
	Dispatcher, Truck 2 is out of service, you can place Rescue 1 back in service That's clear Truck 2, 2111
	Engine 9's in service
	Truck 2 dispatcher, Truck 2 is in service, Engine 2 is out of service, Truck 2 is responding
	Engine 9's in service OK, Engine 9, 2111, Clear
	Paramedic 2 is in service, I'll be enroute to the other scene, 2111
Paramedic 2:	Do you have a medic unit responding over there
Dispatcher:	Not yet, 2111 Clear
	Engine 18 Dispatcher, give us a repeat on the address That will be 3671 Dayton Park Road 18 clear 2111
Dispatcher:	Medic 4 to Dispatch, we can respond

Appendix E (continued)

	Car calling repeat Medic 4 to Dispatch, we can respond to that fire
Dispatcher:	OK, Medic 4 respond to 3671 Dayton Park 2112 4's clear
	West 2's in service 2112
	Truck 2 Dispatcher, we're responding on that, Engine 2 is out of service
Dispatcher:	Engine 8's on the scene, we have a whole building completely involved 2113
	East 2 Dispatcher, I'm on the scene, we've got a large building, about 200 by oh possibly 300, pretty well fully involved. Stage me a third alarm make sure the third alarm response stays out on Wagoner Ford Road. I'll also need dispatch the foam truck, I'll be Sherwin-Williams Command. 2114
Dispatcher:	Engine 9, 11, 16, Truck 16, Truck 15, and the West Chief respond to 3671 Dayton Park Drive, that will be Sherwin Williams, that's a third alarm
Dispatcher:	Engine 9, 11, 16, Truck 16, Truck 15 and the West Chief OK, Truck 15, Engine Truck 16, Engine 9 you clear Clear
Dispatcher:	Engine 11 (responding)--Companies on the scene you're on master You're on Channel 1
Dispatcher:	Engine 11 you clear Clear West Chief West Chief Clear All clear 2116
Dispatcher:	Paramedic 2 to dispatcher, would you have all the medic crews use their telemetry channel 9 to your communication, so they don't mess up our radio traffic out here
	Ok, Paramedic 2
	All Medic companies are you clear on that, use your telemetry instead of the radio channels
East 2:	We'll need one, two medics here on the fire scene, as soon as you can get them here, we do have injuries also, give me rundown of the second alarm response you've got staged out there on Wagoner Ford.
	Truck 11 on the scene 2118

Appendix E (continued)

East 2:	Command to Dispatcher, tell me again what was my second alarm dispatch.
Dispatcher:	OK, command your second alarm response was Engine 14, 4, 2, 18, Truck 2, Truck 11 and the Central Chief
	Chief 3 to dispatcher, we're going to need a lot of traffic control on Wagoner Ford, this is a huge building fully involved, and we have gawkers taking up all of Wagoner Ford 2120
	Command to Dispatcher, quote, listen carefully, I need a run down of the first alarm, second alarm, third alarm companies, give it to me slow so I know exactly what I've got here, what you dispatched.
Dispatcher:	OK, Command, first alarm - Engine 12, 21, 8, Truck 14 Clear so far
Dispatcher:	Standby a second. OK. Command your second alarm Engine 14, 4, 2, 8, Truck 2, Truck 11 and the Central Chief Clear, third alarm
Dispatcher:	Third Alarm - Engine 9, 11, 16, Truck 16 and Truck 15 Chief 1 dispatcher, responding
Dispatcher:	2127
Chief 1:	Are you aware there is somebody at 15's, Engine 15 We are now Medic 15 to the dispatcher, we're in service from Good Sam Do you want us to report to the fire scene
Dispatcher:	Standby at this time, Medic15 Clear standing by Medic 4 removing one to the Valley, burn victim
Dispatcher:	2118 4's is clear 16's in service from St. Elizabeth Engine 16's in the staging area 2129
Chief 3:	Command to Dispatch, this entire structure it is a very large 1 story structure full of flammables. It is completely involved and it will be a total loss. 2132
Chief 3:	As far as possible, I'm going to commit no more resource to the area, our position right now is one of standing by in a very defensive posture 2133 Clear

Appendix E (continued)

Dispatcher:	Engine 2, made an investigation in the 2000 block of Troy Street, we don't know if that's a separate fire or if the people are seeing the fire over on Dayton Park Drive but we're getting a lot of calls about, I have no other information except it's in the, gentleman called from 2050 Troy and he said across the street from him there was a large amount of smoke, your cross street will be Jergens OK, Engine 2, 2135 West
	Medic 8's in service from Good Sam, do you want us to report to the fire scene Ah, standby in the area Medic 8 Clear
Dispatcher:	Dispatcher to Dayton Park Command Go ahead
Dispatcher:	Do you need any medic units over there....Dayton Park Command, were you clear on that? Command to Dispatcher, ah hold that, we have two medic units on the scene that we are aware of, Medic 10 and Box 21 is on the scene, that will be sufficient for the time being Very good, thank you much, we need them
	Inspection 1 is on the air and heading out for the scene 2137
	Dayton Dispatcher, Kettering 4 in service 2137
	Chief 1 on scene
	Engine 2 dispatcher, this call that was called in on Troy Street is part of the large third alarm fire, you can place Engine 2 in service
Dispatcher:	OK, that's what I thought, 2142 Engine 2 clear
	Chief 3 to Dispatch, Chief 3 to all units on the fire scene
	Chief 1 will be Incident Command, Chief 3 will be Operations Command, please address us by those titles, I want all Sector Officer go to Channel 3, are you clear on that dispatch
Dispatcher:	OK, Chief 3, you are Operations That's clear Chief 1 is Chief 1 will be Incident Command
Dispatcher:	Incident, OK, and what was the rest of it then? All Command Officers to Channel 3, Chief 1 will be Dayton Park Command 2144 Clear

Appendix E (continued)

	Investigator 8 Dispatcher, I'm in the area at the scene 2145
	Chief 1 Dispatcher, I'm taking command of the fire, Command Post remains at the northwest corner of the building, this is a 200 by 100' 20' single story building, totally involved probably total loss, will casualties, two we know of, one reported missing. I also have exposures to trailers. Long time on this operation
Dispatcher:	Thank you 2146 Clear
Dispatcher:	Car calling, repeat West 1 dispatcher, what is the address of the third alarm fire 3671 Dayton Park Drive, that runs off Chuck Wagner Lane which is a street that goes into, ah, off Wagoner Ford Road Clear 2149
	Dayton Park Command dispatcher, I'll need police assistance on the scene, we're starting to get infiltration of spectators
Dispatcher:	Police have been notified about this I need them on the scene 2150
	Dispatcher to Dayton Park Command, I informed the Police of the situation and asked for as big as task force as I could possibly muster 2151
	That's clear, we're going to have a tremendous problem with the crowd overrunning this fire scene and it is exploding and still detonating in and around the fire, we cannot guarantee anybody's safety, let along ours
	I've asked for a battalion (interruption by sounding signal) (conversation ended)
	Dayton Dispatcher, Kettering 4 responding to Station 15 That's clear, 2152
	Investigator 1 on the air 2155 Clear
	Dayton dispatcher Kettering 4 out at Station 15
Chief 1:	Dayton Park Command to Dispatcher, can you give me a run down of what we've got in protecting the rest of the city in terms of Chief Officers and equipment
Dispatcher:	Ok, at this time we have Engine 10, we have at Co. 2's, 15's at Co. 4's, Madison truck and a Moraine engine at Co. 11's

Appendix E (continued)

	We got Engine and Truck 13 in quarters, Harrison Township has a truck and an engine at Co 14's, Kettering engine and truck at 15's and Engine 17. We're well covered as far as the city's concerned.
Chief 1:	Do you have Chief Officers in three districts?
Dispatcher:	There's some enroute now
Chief 1:	Take your first 3 chief officers that report on the air and assign them districts, send the next chief officers to this fire. I do not need chief officer support more than I need them to protect the city
Dispatcher:	That's clear, we have Chief 4 at Headquarters also
	Dispatcher to East 1, Dispatcher West 1
East 1:	Go ahead, East 1
Dispatcher:	Yeah, are you enroute to quarters now I'm enroute to 2's to pick up a driver
Dispatcher:	That's clear, then you'll take the East side Chief then East 1 clear
	Dispatcher to West 1 West 1, I've been given assignment by Operations Command I'm nearly on the scene of the fire now
Chief 1	Cancel that, this is Incident Command, go run the district We'll get you something else
Dispatcher:	West 1 you'll be West Side Chief West 1 clear 2159
	Investigator 4 on scene 2204
	Dispatcher to Incident Command, have any removals been made to area hospitals
Dispatcher:	We have an indication that we have 1 removal but I don't know where the individual was removed to That's clear 2204
	Medic 4 to Dispatch that removal was made by us to the Valley 2205 4's is in service and clear
Chief 1:	Command to Dispatcher, has the City Manager been notified of this incident?

Appendix E (continued)

Dispatcher:	Could you repeat your message please Has the City Manager been notified of this incident? We're notifying him now Thank you 2205
	Dispatcher to Incident Command
Chief 1:	Command go ahead
Dispatcher:	We're getting a lot of calls, has there been any talk about evacuation out there
Chief 1:	No, evacuation at this point is not necessary, we have a fire that is essentially isolated with the exception of an exposure of trailers around the building, the closest other structures are at least 100 feet away, there is a considerable amount of exploding and evolving fire, a this point however, does not need evacuation
Dispatcher:	That's clear, 2209
Dispatcher:	Chief 1, the Assistant Manager is concerned with the reclamation fields out there as far as water tables, is there any concern there, does he need to contact anybody?
Chief 1:	We have made contact with the Water Department, they're enroute and we also have the Environmental Specialist from the Water Department on scene monitoring that situation You can assure him that we're taking a close look at it and we'll make whatever decisions we need to secure the water system
Dispatcher:	He requested that we contact him if there is any significant changes, so if you want to relay that through us, we can handle it.
Dispatcher:	Dispatcher to Incident Command Command go ahead
Dispatcher:	Yes sir, we have the names of six employees from out there and when this incident occurred they fled the scene, but we do have six names here if you need any
Chief 1:	I would like to confirm if all six are able to talk to you, we've made one removal, if you can talk to the other five or talk to someone who has, I'd like to know that, we're treating it as though we have victims
Dispatcher:	OK, each one of these people on the list have been confirmed
Chief 1:	So, in other words, all six have been confirmed as alive
Dispatcher:	That is correct, 2212 One injury
Dispatcher:	We've have Medic call us when they get back with the injury and get a report Thank you 2212

Appendix E (continued)

	Investigator 1 Dispatcher why don't you contact Investigator 7 have them stop by the Valley and let him talk to that person
Dispatcher:	2213, Investigator 7 Go ahead
Dispatcher:	Would you stop by the Valley and talk to the person removed from Wagoner Ford I couldn't copy you
Dispatcher:	Would you stop by Miami Valley Hospital and talk to the individual who was removed from Wagoner Ford Road That's clear sir
Dispatcher:	That's per Investigator 1 at 2213
	Command to Dispatcher, could you confirm whether or not you talked to Joe Crone
Dispatcher:	No, we that's not one of the names we have sir That's the one we're hunting 2213
	Inspector 4 in service 2213 Inspector 4 clear
	Training Center responding to the fire on Dayton Park 2214 Clear
Chief 1	Command to Dispatcher, were you able to contact the Public Information Officer
Dispatcher:	We're working on that now Thank you
	Dispatcher Dayton Fire Command
Chief 1:	Command go ahead Do you have a Medical Sector there, we need to know what medics we have on the scene
Chief 1:	I'll get back with you in just a second, we do have Chief 5 yes, Chief 5 is Medical Sector 2217
	Investigator 1 is on the scene 2218

Appendix E (continued)

	Command to Dispatcher, we have Medic 4 and Medic 10 along with Harrison Township Medic, and a Box 21 Unit assigned to this, make that Medic 8 and Medic 10
Dispatcher:	Thank you, 2218 That's clear, we'll hold those by the way
Dispatcher:	That's clear, 2218
Chief 1:	Command to Dispatcher, do you have any other day off chief officers responding to this fire
Dispatcher:	That's negative None available
Dispatcher:	We have them in the district we can send out I'm asking you were they contacted and they did not respond
Dispatcher:	Command, every chief we got a hold of is responding
Chief 1:	I can't hear you
Dispatcher:	Every chief we got a hold of is responding
Chief 1:	That's clear, now I'm going to ask you again, do you have enough that your districts are filled and a surplus of chief officers are coming to the fire scene
Dispatcher:	We have no surplus at this time That's clear
Dispatcher:	The districts are covered All three districts are covered
Dispatcher:	Affirmative Thank you 2220 Investigator 7 I'll be at the Valley 2223
Chief 1:	Command to Dispatcher can we get some kind of a canteen out here, we've got some people that are getting pretty thirsty and dehydrated, Box 21, Red Cross, somebody would help
Dispatcher:	I believe Red Cross is enroute and Box 21 also
Chief 1:	That's clear Command to Dispatcher, have you been able to contact anybody from the Training Center They're on their way out there
Chief 1	Clear, thank you 2227

Appendix E (continued)

Investigator 5 on the scene
2232

Dispatcher: Dispatcher to Dayton Park Command

Chief 1:   Command go ahead
The man that you were looking for, Joe Crone, he is home

Chief 1:   Clear, thank you
230

Garage 4 Dispatch
Go ahead
Garage and Garage 2 are on their way to the scene on the air
2234

Maintenance 3 Dispatch
Go ahead
Maintenance 3 and 4 we're enroute to the scene with fuel

Dispatcher: Would you repeat your message please
Maintenance 3 and 4 enroute to the scene with diesel fuel
2237

Maintenance 3:   Do you go off Wagoner Ford Road
Would you repeat your message
Is the location off Wagoner Ford Road

Dispatcher: At Dayton Park Drive
Clear
2237

Dispatcher: Incident Command, have the site management people from that company contacted you at the scene
Yes they have

Dispatcher: Clear they have a liaison established at the Radisson
They were contacting us to make sure you had people on the scene
Yes plant manager's been here

Dayton Park Command to Dispatcher, can you give me a rundown on the time of alarm, and calls for equipment

Dispatcher: OK, Incident Command at 2107 we sent the first alarm at 2110 we sent a second alarm response, and then at 2116 a third alarm response was sent

Chief 1:   That's clear, thank you
2249

Command to Dispatcher, has EPA been notified of this fire
Not at this time

Appendix E (continued)

Chief 1:	Ah, call their office and at least make them aware of what's happening 2253 Clear
	Dispatcher to Incident Command Go ahead
Dispatcher:	Did you say you did have a fatality out there
Chief 1:	We are unable to determine that at this time Ok, so you won't be needing the chaplains or anything out there We will notify you
Dispatcher:	That's clear 2255 Dispatcher to Incident Command, EPA has been aware of the situation
Chief 1:	That's clear 2257
	Investigator 7 cleared from Miami Valley enroute to the scene 2304
	Medic 14 Dispatcher Go ahead We're in service with the medic 2306
	Command to Dispatcher, I understand you had an inquiry about the need for a chaplain Yes we did
Chief 1:	I don't think we'll need them for anybody in the normal sense here, we believe we've got everybody accounted for ah, and short of some kind of injury to personnel here or breakdown by management here, I don't think we'll need any of the chaplains assistance at this point
Dispatcher:	2316 Dispatcher to Incident Command
Chief 1:	Incident Command Dispatcher, go ahead
Dispatcher:	Ok, Wright Pat called us on the phone and said if there is a need for foam that they do have a foam truck available
Chief 1:	That's clear, we'll keep that in mind as a resource, we're still trying to determine whether or not we should make an application of an extinguishing agent because of the possible contamination of the wellfield 2327
Chief 1:	Thank you

Appendix E (continued)

	Medic 18 Dayton Dispatcher, Medic 18 will be in service Medic 18 2328 Medic 18 clear
Dispatcher:	Medic 18 come down to fill in at 4's Clear 2328
Dispatcher:	Medic 18 fill in at 13's Clear 2329
	Medic 18, Dispatcher to Medic 18 Go ahead Are you a two person crew right now Affirmative Medic 18, go by Co. 14's pick up 1 paramedic there and you'll be running with 3 paramedics That's clear we'll be enroute to Co. 14
Dispatcher:	That's clear 2333 Inspection 3 enroute to the fire scene
Dispatcher:	Maintenance 3 did you say you're enroute to the fire scene Inspection 3, that is correct
Dispatcher:	Maintenance 3 you're breaking up cannot copy Inspection 3 is enroute to the fire scene
Dispatcher:	That's Inspection 3? That's correct 2335
	Investigator 1 to Dispatcher, do you have a location as to where the employees went to meet
Dispatcher:	They've been calling in from all over They didn't go to meet at one place or location
Dispatcher:	Not that I know of Clear
Dispatcher:	2336
	Dispatcher to all companies, if you have all unassigned personnel in quarters, call 3316
	That's dispatcher to all companies, if you have any unassigned personnel in quarters at this time, call 3316. 2340

Appendix E (continued)

Dispatcher to Investigator 1
Investigator 1 Dispatch
Go ahead
You called me
Yes sir, we have a report that a few of the employees from there are up at the restaurant on Webster and Wagoner Ford
At Webster and Wagoner Ford
That's correct
Clear thank you
2342

Medic 18 to Dispatch, we have our third person enroute to Co. 13's
2347

Dayton Dispatcher, Kettering 4 leaving Station 15 enroute to quarters

Dispatcher: Thank you, Kettering 2347

May 28, 1987

Dayton Park Command to Dispatcher, this fire is essentially confined, not under control, we will probably have crews here throughout the night, but I do want to indicate that it is no longer in such a mode that it is continuing to expand, but what we got will burn for a considerable length of time

Dispatcher: Thank you command 0012

Thank you

Dispatcher to Incident Command

Chief 1:   Command go ahead

Dispatcher: Will you be needing anymore Garage personnel out there, they have 3 at the Garage at this time

Chief 1:   I don't know, we're going to have a fueling problem that kind of problem over a long hall operation, these crews will be here all night and probably into tomorrow.

Dispatcher   0018

Command to Dispatcher, would you have the Central Chief get his explosive meter and run it out here, if he's not sure as to the operating condition of the explosive meter, poll the chief officers on duty, get one that works, get it out to me as quick as we can, please.

Dispatcher: 0038
Clear

Dispatcher to Incident Command

Appendix E (continued)

Chief 1:	Clear Ok, Central Chief is on his way to 11's to pickup that explosive meter, they'll be enroute to your location That's clear 0045
	Command to Dispatcher, can you give me the time that I indicated this fire was confined
Dispatcher:	That time will be 0012 Thank you
Dispatcher	052
	Dispatcher to Incident Command
Chief 1:	Incident Command, go ahead Have you got a mechanic there that you can release to go to Miami Valley Hospital, Medic 15 will not start
Chief 1:	We got a mechanic running around here, we'll relay the message 134
	Dispatch, this is Garage 4
Dispatcher:	Go ahead Ah, we're enroute to pick up some more diesel, you want us to swing by the Valley and check it out
Dispatcher:	Yes sir you can do that, you going to go before you get the diesel? Ah yeah, we can stop by before we fill up
	Dispatcher to Incident Command Command, go ahead Do you have an urgent need for diesel fuel at the scene there
Chief 1:	I haven't been made aware of any
Dispatcher:	That's clear, 135
Dispatcher:	Ok, Garage 4 continue on to the hospital Ah, that's a roger, we're enroute right now
Dispatcher:	135 Dispatcher to Medic15 15's go ahead
Dispatcher:	Garage 4 is enroute to your location at this time Clear 136

Appendix E (continued)

	Incident Command, Dispatcher, would you send us your closest engine with an AP and we'll release one of your staged engines, we've got some brush fires that we need to deal with out here, we'll need a four wheel drive vehicle
Dispatcher:	156 Clear
Dispatcher:	What do you actually have in staging at this time?
Chief 1:	We have Engines 9, 11, 16, Truck 15 and Truck 16 and Engine 18 We'll release Engine 18
Dispatcher:	That is clear
Dispatcher:	Engine and AP 15 respond to the fire scene, that will be 3671 Dayton Park Drive
	That's Engine and AP 15 respond to 3671 Dayton Park Drive That's at the fire scene Ok, Engine 15 and AP 15 158 Coleman
	Engine 18 Dispatcher, we're in service from the fire scene enroute to quarters
Chief 1:	Command to Dispatcher, can you tell me what we've got staffing our stations right now and where they're located
Dispatcher:	Co. 2's we have Engine 10; Co. 4's we have Engine 110 and Medic 4; Co. 8's we have Engine 108, Co. 11's we have Engine 111 and Truck 111; Co. 12's we have Huber Heights Reserve and standby personnel; Truck 13 at 13's, Engine 13 at 13's; Engine 116, Truck 113 at Co. 14; Engine 118 at Co.'s 15; Engine and AP 17 at Co 17's; Engine 18 enroute to Co. 18's; Engine 2, Truck 102 at the fire scene at 115 Samuel; that's it for now.
Chief 1:	Thank you, it looks like we're still in pretty good shape
Dispatcher:	That's true, and we do have a chief for each district at this time
Chief 1:	Thank you 206
	Engine 15's on the fire scene 214
	Investigator 1 Dispatcher, all investigators have been released from the fire scene 217 Clear
	Investigator 1 Dispatcher, when Investigator 6 clears the scene on Samuel would you inform him that you can send him home
Dispatcher:	That's clear, Investigator 6 are you clear on that message?

Appendix E (continued)

Dispatcher to Investigator 6
228

Investigator 6 clear of Samuel, and out to the fire scene

Dispatcher: Ah, per Investigator 1, you are released to go home
Ok, Investigator 6 is clear, thank you
236

Investigator 7 I'll be off the air
241

Medic 10 dispatcher we're released by Incident Command we'll be remaining out of service, going to the Valley for equipment
256
Medic 10's clear

Chief 1:   Incident Commander to Dispatcher, I'm going to be releasing Engine 9, 11, 16, Truck 15, Truck 16 and Truck 2 and you can release all day-off personnel when they go in service

Dispatcher: That's clear, 257

Chief 1:   For all intensive purposes, this fire will continue to burn throughout the night into the morning, part of the day tomorrow I guess, we're gong to continue to let this burn and the companies that are on scene will be evaluated as to what will be kept and released when I shift command over.

Dispatcher: 257
Clear

Truck 16's in service from the fire scene
302
Truck 16 clear

Truck 2's in service from the fire scene
303
Truck 2 clear

Paramedic 2 in service
303
Paramedic 2 clear

Truck 15 in service leaving the fire scene
303
15 clear

Engine 9's in service
303
Clear

Appendix E (continued)

Chief 1:   Command to Dispatcher, command of this incident is being transferred to Chief 3, Chief 1 will be in service, these crews, again, will be here most of the night, in fact, all night.

Dispatcher: 304

(Tape Transcription Ended from Dispatcher Log Tape)

Dayton Daily News and Journal Herald   May 29, 1987

'Environmental nightmare come true'

By Jim Babcock
STAFF WRITER

For Sierra Club leader Joe Bockelman, the fire-leveled Sherwin-Williams Automotive Distribution Center had long been "a problem waiting to be an accident."

For Dayton City Commissioner Mark Henry, the raging, chemical-fed conflagration at the paint warehouse was "an environmental nightmare come true."

What worried both men, and the Ohio Environmental Protection Agency as well, was the building's proximity to Dayton's 830-acre Miami River Well Field.

"The whole situation out there is one that never should have happened," Bockelman, who is vice chair of the local Tecumseh Group of the Sierra Club, said Thursday.

"I would point to that as an example that well fields and industrial parks are two beasts that do not belong together — especially where the geologic setting affords no protection to our ground water."

The charred, still-smoldering remains of the Sherwin-Williams warehouse are near the center of an 84-acre site known as the Concourse 70/75 Industrial Park.

The city-owned site was opened to industrial development in 1973, and at one time was envisioned as the core of an industrial park that would spread over much of the Miami well field — which presently is occupied by the Kittyhawk Golf Course, a system of city water department recharge lagoons and a water-treatment plant and pumping station.

"But the way that City Hall weighed the policy options then, I guess, was a little different than we probably would now,"

said Henry, who was elected to the City Commission in 1983.

"It would be my position now that that site wouldn't be used for any other industry — unless they want to propose a cotton-candy warehouse or something like that."

The city changed its mind about expanding Concourse park in 1983, after a Florida consulting firm warned that further development would greatly increase "risks in destroying the water resource" underlying the site and the adjacent well field.

The consultant, CH2M Hill, also warned that "a significant potential hazard" already existed at five of about 20 structures on the site. And the company recommended a series of steps to prevent the release of hazardous substances — including keeping-

SEE ENVIRONMENT/4

Dots show locations of wells

WAGONER FORD RD.
BEARDSHEAR RD.
Water treatment plant
Recharge lagoons
DAYTON PARK DRIVE
Site of fire
Great Miami River
B&O RAILROAD

KEVIN RILEY/STAFF

SOURCE: City of Dayton

Appendix F (continued)

4 ••   **Dayton Daily News and Journal Herald**   Fri., May 29, 1987

☐ Environment

CONTINUED FROM/1

out "prospective tenants which use or have as a byproduct . . . any material listed or classified as hazardous."

The CH2M Hill recommendations led to a June 22, 1983, City Commission resolution promising that Concourse park would not be expanded and that all future development within the park would be carefully screened to assure that it posed minimum potential for contaminating ground water.

But that step failed to satisfy the Ohio EPA, which had specifically cited the Sherwin-Williams warehouse as a high risk in an October 1980 letter asking the city to take steps to minimize the potential for ground water contamination from chemical spills.

The letter noted that the warehouse contained an estimated 1.5 million gallons of paint and paint solvents that could cause extensive contamination if they were accidentally spilled.

Ironically, city and Sherwin-Williams officials agreed that a "major fire" at the warehouse could pose the greatest hazard if paints and solvents were carried onto the wellfield with water from firefighters' hoses.

But the officials also said fire was a remote possibility, because the warehouse was equipped with an automatic sprinkler system. And the Sherwin-Williams spokesman surmised that if there were a fire, most of the chemicals would be consumed by flames.

Fire officials said Thursday, though, that the fire was so intense because of those chemicals that it overwhelmed the sprinkler system.

After speaking out in 1980, the Ohio EPA did not again make its concerns public until early 1983, when it criticized the city for permitting another Concourse tenant, Purolator Courier, to bury two 12,000-gallon fuel tanks at the company's new location in the park.

Then in 1984 — after traces of cancer-causing industrial degreasing solvents were detected in several wells near the north end of Kittyhawk Golf Course — the state agency began intensified negotiations that led to an agreement committing the city to refrain from developing any remaining vacant parcels in the industrial park.

The January 1985 agreement also committed the city to development of a well field protection and management plan and to requiring Concourse's existing tenants to establish safeguards to prevent chemical and fuel spills from escaping to areas where they could seep into ground water.

Henry said he feels the agreement has helped establish "an uneasy middle ground."

"The tough question we all reach at some point in time is, what do we do about business development that already exists in environmentally sensitive areas? . . . You make it as safe as possible. You take all the steps you can take and put in safeguards so that if things do happen, you don't get hurt as bad," Henry said.

Henry also said the Sherwin-Williams warehouse "was the greatest concern" because of "the sheer quantity of the contaminants they handled."

But he surmised that safeguards already installed in the warehouse may have helped contain unburned chemicals.

"For example, a concrete apron they were required to extend probably helped keep runoff from spreading," he said.

Bockelman saw the situation differently, however. The Sierra Club official charged that the city has been slow in implementing the wellfield protection steps called for by the agreement.

"The city has been playing this game too long — kind of toying with the risks," he said.

"They should have tried to relocate Sherwin-Williams right away. It just flies in the face of modern environmental protection. . . . These water problems we're facing are just textbook illustrations of what government is supposed to prevent."

Bockelman also said his criticisms of Dayton's wellfield protection efforts extend as well to the city's Mad River Well Field, which stretches between Finley Street and Rohrer's Island in East Dayton.

"I guess this has brought us to the point where we feel we've got to abandon the (Concourse) industrial park. The water is just too delicate a resource to jeopardize with facilities such as Sherwin-Williams. And if that's true, we must abandon Gateway (Industrial Park, near the Mad River field) as well.

"But we've committed millions of dollars," Henry said, "and the city is trying to implement one of the most progressive water management plans in the country."